Contents

Competition and
Coexistence of Species

Competition and Coexistence of Species

A J Pontin
Royal Holloway College
University of London

Pitman Advanced Publishing Program
BOSTON · LONDON · MELBOURNE

PITMAN BOOKS LIMITED
39 Parker Street, London WC2B 5PB

PITMAN PUBLISHING INC
1020 Plain Street, Marshfield, Massachusetts

Associated Companies
Pitman Publishing Pty Ltd, Melbourne
Pitman Publishing New Zealand Ltd, Wellington
Copp Clark Pitman, Toronto

Library of Congress Cataloging in Publication Data
Pontin, A. J.
  Competition and coexistence of species.
  Includes index.
  1. Competition (Biology) I. Title. II. Title:
Coexistence of species.
QH546.3.P66   574.5   81-1654
ISBN 0-273-08489-5   AACR2

British Library Cataloguing in Publication Data
Pontin, A. J.
  Competition and coexistence of species.
  1. Competition in ecology
  I. Title
  574.5'24   QH546.3
ISBN 0-273-08489-5

Text set in 10/12 pt Linotron 202 Times, printed and bound
in Great Britain at The Pitman Press, Bath

Preface

In spite of the outpouring of papers on the many uncoordinated facets of interspecies competition there has been no recent monograph which attempts a general synthesis for students to read. All accounts are strongly biased and this book is no exception, but there seemed little point in writing an uncritical review so I offer it as further fuel for controversy. However, the bias here is towards factual evidence which is the least misleading basis for argument and which deserves to be more widely read.

The rate of publication of papers is a major problem and it was found impossible to carry on reading many of them during 1980 and also find time for writing. Without the Biological Information Service provided by UKCIS, University of Nottingham, the task would have been even more difficult and I wish to acknowledge the valuable support given me by Royal Holloway College Library in this. I also wish to thank the many authors and publishers who gave their permission for me to include their work.

A. J. P.

1 Introduction

1.1 General background

Variety, not monotony, is a normal feature of ecological communities. It is exceptional for there to be only one photosynthetic plant species or only one species of herbivore or even only one species of carnivore in any area. So there are two main components of ecological variety: firstly food chains with a limited number of links which might be exemplified by oak leaves, insects, passerine birds and sparrowhawks and, secondly, a much less restricted number of potentially competing species sharing the food at each level – there are hundreds of insect species able to feed on oak. This book is concerned with the second sort of interaction or, in other words, with processes leading to energy sharing rather than energy flow and why the share-out is so complex.

Darwin drew attention to the importance of competition for resources (food or other necessities of life) which are not adequate for the survival of all the individuals born. A quotation from the 'Origin of Species' (1859, p. 396) summarizes his views on the subject: 'More individuals are born than can possibly survive. A grain in the balance will determine which individuals shall live and which shall die, – which variety or species shall increase in number, and which shall decrease, or finally become extinct. As the individuals of the same species come in all respects into the closest competition with each other, the struggle will generally be most severe between them; it will be almost equally severe between varieties of the same species, and next in severity between the species of the same genus. But the struggle will often be very severe between beings most remote in the scale of nature. The slightest advantage in one being, at any age or during any season, over those with which it comes into competition, or better adaptation in however slight a degree to the surrounding physical conditions, will turn the balance.'

This point of view led to concentration of effort on discovering examples of one species replacing another and an unfortunate neglect of coexisting species. For example, Gause (1934) in his book 'Struggle for Existence' makes it clear that his aim is to investigate Darwin's question: 'Why is one species victorious over another in the great battle of life?' The analogy between two competing species and two armies in a battle is false and has been particularly misleading. Obviously any slight advantage maintained by one side in a battle will 'turn the balance' and lead to the defeat of the other side. Continuous battle between two equally matched armies is not to be expected, but interspecific competition is quite different. The really fundamental difference is that intraspecific competition must be going on as well as interspecific competition if resources are in short supply. 'Soldiers' on the same side are greater rivals than 'soldiers' on opposite sides when the opposing sides are two species competing for a share of food. Darwin pointed this out, but the last sentence of his quoted argument does not follow from these premises. There is also a potentially confusing uncertainty about whether Darwin is talking about individuals or populations or whole species. To try and clarify these distinctions, Chapters 2 and 3 will deal with behaviour of individuals, Chapters 4 and 5 with populations and Chapters 6 and 7 will attempt a broader view of species.

Lotka (1925) and Volterra (1926, translated in Chapman, 1931) put forward a mathematical analogy for interspecific competition which has become standard theory for discussion of the subject. The predicted results of competition between two species may be any of the possible alternatives, i.e. (1) species A always displaces B, (2) instability with displacement of either by the other depending upon starting density or other factors, or (3) stable coexistence. This third possibility results from each species limiting its own increase more than that of the other, and coupling this concept with Darwin's axiom that intraspecies competition is generally more severe than interspecies competition, it can be argued that stable coexistence with continuous competition is of frequent occurrence (Pontin, 1958, 1961; Cole, 1960).

It is wrong to make the common assumption (e.g. Lack, 1971) that two coexisting species are ecologically isolated or controlled by separate population-limiting factors and this argument is enlarged in Chapter 4. The corollary that competitors cannot coexist is factually incorrect and in any case should not have been accepted after the publications of Lotka and Volterra.

All species can be expected to differ from one another and, where sufficient research has been done (e.g. *Tribolium*, Lerner and Ho, 1961; or *Drosophila*, Mather and Cooke, 1962; Seaton and Antonovics, 1967), the different genotypes of one species show differences in competitive performance. The competitive exclusion principle (Hardin, 1960) which states that 'complete' competitors cannot coexist is therefore not worth saying. Complete competitors cannot be expected to exist at all (Pontin, 1963).

This leads us to the repeatedly asked question: How similar can two species be before coexistence is impossible? This rather unrealistic idea is discussed further in Chapter 6 and examples of extreme similarity of competing species are given in Chapter 4.

1.2 Definitions

Interspecific competition is not easy to define. When it meant simple displacement of one species by another the term was used loosely (Birch, 1957) and the need for a more precise definition hardly arose, but as soon as competition with stable coexistence is admitted to be possible then a definition becomes necessary for experimental diagnosis of supposed interactions. There are almost as many definitions as authors and many imply (e.g. Milne, 1961) that competition is the behaviour observed when individuals interact although this may or may not have any effect on their populations which would be necessary to make it ecologically interesting. One of the few authors presenting a clear ecological meaning for competition is Odum (1953) and he classifies the possible interactions between two species by their effects on 'population growth and survival' (Table 1.1).

Table 1.1 The main types of direct interaction between two species

	Effect on sp. 1	Effect on sp. 2
Herbivore/plant, predator/prey, parasite/host	+	−
Commensalism	+	0
Symbiosis	+	+
Competition	−	−
Amensalism	−	0

Interspecific competition is characterized by a negative effect on both species competing and it would be better to specify this effect more precisely as a reduction in population size (Pontin, 1961). Measurement of population size by density of individuals is not always practicable and difficulties arise from three main causes: (1) individuals may be difficult to assess because they are colonial or spread vegetatively, (2) individual size may be very variable or plastic – flatworms can even de-grow in response to starvation, or (3) the species may be so different that comparison of numbers is not meaningful and the competition between seed-eating rodents and ants (Chapter 4) provides an example. It may be necessary to measure biomass or even metabolism in some cases and measurement of fecundity has frequently been found a sensitive way to assess interaction. Whatever population measurement is used the basic approach should be to remove experimentally one species from its normal habitat and to look for increase in population size of the remaining species compared with controls. Reciprocal effects should also be looked for to satisfy Odum's definition. Without this type of investigation the ecological effects of competition are hypothetical.

The definition of interspecific competition used here is *an interaction between two (or more) species which results in reduced population size of both (or all) competing species.* Any ecologist who insists that competition has a different meaning will need to invent a new term for this ecologically important effect.

The term 'stable' is used here to describe *a system which automatically tends to return to an equilibrium state after disturbance.*

Two types of behaviour may cause competition (Park, 1954; Elton and Miller, 1954; Brian, 1956) and these are: (1) 'exploitation' where *individuals of different species are observed to use some of the same resources*, and (2) 'interference' where *fighting or other directly damaging behaviour* occurs.

Observations of such behaviour alone do not, of course, enable one to conclude that the species are affecting each others' population sizes. Two insect species may eat the same food-plant species, but their numbers may theoretically be kept so low by parasites and predators that food sharing is insignificant in effect. Such superabundance of resources can be deceptive because they could be selecting a specific age of leaf and the question of population interaction needs direct measurement. There are many published papers which refer to behaviour as competition because the authors are using a different meaning for the term and these will not be included in Chapters 4 and 5 as examples of competition.

Interference and exploitation need to be distinguished because they can be expected to have different consequences at the population level (Pontin, 1963; Gill, 1972; Conley, 1976). Exploitation without any interference is likely to produce stable competition because intraspecific competition is automatically likely to exceed interspecific competition in severity (*see* Chapter 4 for another way of expressing this argument). However, in the case of interference, individuals of the same species may well be evenly matched, but individuals of different species may be ill-matched with one species clearly dominant over the other. The dominant species is likely to displace the other if interference effects override the basically stabilizing effects of interspecific exploitation. This distinction is most clearly appreciated if man is one of the competing species. Imagine trying to eliminate rabbits or locusts by superior exploitation of the vegetation. Man's attempts to displace pests are interference methods which are not expected to affect his own species density by killing people or reducing their fecundity.

References

Birch, L. C. (1957). The meanings of competition. *Amer. Naturalist* **91,** 5–18.

Brian, M. V. (1956). Exploitation and interference in interspecies competition. *J. Anim. Ecol.* **25,** 339–46.

Chapman, R. (1931). *Animal Ecology.* New York and London: McGraw–Hill.

Cole, L. C. (1960). Competitive exclusion. *Science* **132,** 348–9.

Conley, W. (1976). Competition between *Microtus*, a behavioural hypothesis. *Ecology* **57,** 224–37.

Darwin, C. (1859). *The Origin of Species by Means of Natural Selection.* Harvard Facsimile 1st edition, 1964.

Elton, C. S. and Miller, R. S. (1954). The ecological survey of animal communities: with a practical system of classifying habitats by structural characters. *J. Ecol.* **42,** 460–96.

Gause, G. F. (1934). *The Struggle for Existence*. Baltimore: Williams and Wilkins Co.

Gill, D. E. (1972). Intrinsic rates of increase, saturation densities and competitive ability. Part 1 an experiment with *Paramecium*. *Amer. Naturalist* **106**, 461–71.

Hardin, G. (1960). The competitive exclusion principle. *Science* **131**, 1292–7.

Lack, D. (1971). *Ecological Isolation in Birds*. Oxford: Blackwell.

Lerner, I. M. and Ho, F. K. (1961). Genotype and competitive ability in *Tribolium* species. *Amer. Naturalist* **95**, 329–43.

Lotka, A. J. (1925). *Elements of Physical Biology*. Baltimore. Hopkins & Williams.

Mather, K. and Cooke, P. (1962). Differences in competitive ability between genotypes of *Drosophila*. *Heredity* **17**, 381–407.

Milne, A. (1961). Definition of competition among animals. *Symp. Soc. Exp. Biol.* **15**, 40–61.

Odum, E. P. (1953). *Fundamentals of Ecology*. Philadelphia: Saunders.

Park, T. (1954). Experimental studies of interspecific competition: II. Temperature, humidity, and competition in two species of *Tribolium*. *Physiol. Zool.* **27**, 177–238.

Pontin, A. J. (1958). *Colony Foundation and Competition Between Ants*. D.Phil. thesis, University of Oxford.

Pontin, A. J. (1961). Population stabilization and competition between the ants *Lasius flavus* (F.) and *L. niger* (L.). *J. Anim. Ecol.* **30**, 47–54.

Pontin, A. J. (1963). Further considerations of competition and the ecology of the ants *Lasius flavus* (F.) and *L. niger* (L.). *J. Anim. Ecol.* **32**, 565–74.

Seaton, A. J. P. and Antonovics, J. (1967). Population interrelationships. I. Evolution in mixtures of *Drosophila* mutants. *Heredity* **22**, 19–33.

Volterra, V. (1926). Variazioni e fluttuazioni del numero d'individui in specie animali conviventi. *Mem. R. Acad. Naz. dei Lincei* **2**, 31–113.

2 The causes of competition: exploitation

The behaviour which results in the sharing of resources between two or more species may be classified as follows: (a) Specialization by selection of qualitatively different resources. (b) Chance acquisition of particulate resources without preselection. (c) Stratification of the species into different zones of space or time. (d) Interspecific territory producing a mosaic (*see* Chapter 3).

2.1 Specialization

It is amazing that many animals requiring the same basic food substances, which are available from a wide range of food species, select only a small proportion of that which appears suitable and starve rather than eat alternatives. There are most examples among herbivorous insects or parasites, but any biologists can think of examples in their own field of interest.

In England the chalk hill blue butterfly (*Lysandra corydon*) appears to be restricted in range by its foodplant, *Hippocrepis comosa*, which is confined to chalk and limestone grassland (*see* Ford, 1945). It is numerous locally, but related species which eat a wide variety of leguminous plants, e.g. the common blue (*Polyommatus icarus*), are much more widespread and must be much more numerous overall.

Most lepidoptera are not as specialized as this and it is intriguing to compare their taxonomy with ours. Many specialize on a genus of plants, for example the fritillary butterflies on *Viola*, and the genera of trees are more useful than species for correlating British lepidoptera with their food plants. Many pairs of genera seem to be more similar to phytophagous insects than to us. These include *Epilobium* (Onagraceae) and *Galium* (Rubiaceae) to the elephant hawkmoths, *Deilephila*, or, *Brassica* (Cruciferae) and *Tropaeolum majus* (Tropaeolaceae) to the cabbage white butterflies *Pieris brassicae* and *P. rapae*. There are many more inconsistencies, but if we based our taxonomy on plant secondary chemicals we might understand better. It certainly appears (*see* Ehrlich and Raven, 1964) that foodplant selection is a combination of attraction by specific chemicals, which may have been evolved as plant defences in the first place, and repulsion by chemicals from the wrong plants.

The other phytophagous insect groups show parallel examples and some-times specialization is even more extreme. Janzen (1977) working on Costa Rican seed-eating beetles found 95 species on 83 species of tree; 78 bruchids had only one host and 17 cerambicids and curculionids also were specific to single-tree species.

There must be strong reasons why specialization is commonly evolved and maintained in spite of the loss of potential resources and reduction in maximum population size attainable. Four types of explanation have been put forward: (a) avoidance of competitors, (b) increased efficiency, (c) coevolu-tion, and (d) apostatic selection. It is worth discussing each of these in turn, but some of them are only side-issues here and will be dealt with briefly.

Competitor avoidance is superficially plausible as an explanation, but it does not bear closer examination. It is actually unusual to find a specialist which is the only species exploiting a particular food. Several species of insect are quite frequently found on the same foodplant and furthermore they are often closely related. The leafhoppers of genus *Erythroneura* described by Ross (1957) (*see* Chapter 4) are a good example with up to seven species occurring on the same tree. The term 'guild' has been used (Root, 1967) for such assemblages of species. Divergent specialization of two species which at the same time increases their efficiencies is much more likely to be a realistic explanation of some cases.

The second type of explanation usually claims that morphological adapta-tions for finding and assimilating one sort of food render the species less well adapted to different food. The bill shapes of birds are probable illustrations of this and obviously one would not expect a duck to feed on floral nectar. Darwin's finches (Lack, 1947; Abbott, Abbott and Grant, 1977) provide a well-documented example which shows on a small scale what may have been a general phenomenon in bird radiation. The overlap in seed and fruits eaten by *Geospiza* species is inversely related to their difference in beak depth and where only two species occur on small islands of the Galapagos group the pair tends to be very different in beak depth (Fig. 2.1). Evolution of specialization of beak and food by competing species certainly seems the likely explanation.

Similarly it might be more efficient for an insect larva to evolve a physiological adaptation for countering the toxins of one foodplant species rather than possess a larger range of such adaptations when only one is needed in the life of any one individual. However, examples of extreme polyphagy do exist and several species of British moths will eat an enormous variety of broadleaved tree when larvae. One of these, the lackey moth *Malacosoma neustria* will eat the cyanogenic cherry laurel (*Prunus lauro-cerasus*) as one of its many alternative foods.

Coevolution of eater and eaten may also be a source of diversity in some groups. Put simply, the food species have radiated out from a common ancestor with allopatric speciation and their dependent eaters followed the same spatial pattern with parallel speciation. The closely matching classifica-

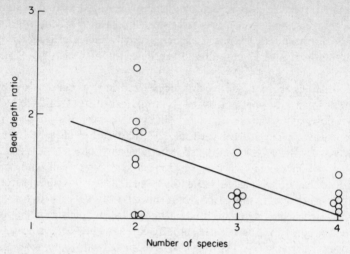

Figure 2.1   Ratio of beak depth of most similar sympatric pair of *Geopspiza* species on each island of the Galapagos group (ordinate) plotted against the number of sympatric species. (*From* Abbott, Abbott and Grant, 1977.)

tions of birds and their parasites (Rothschild and Clay, 1952) have been interpreted in this way to the point where bird classification is thought to be confirmed or doubted because of their parasites. For example most orders of birds have their specific genera of lice except for the woodpeckers all of whose louse species occur on passerines suggesting that they are also passerines. The grebes and divers have different characteristic genera of birdlice and also quite different tapeworms which would lend weight to separation of them into unrelated orders. On the other hand ostrich and rhea have a distinct and specific genus of louse *Struthiolipeurus* in common and also the same tapeworm *Houttuynia* which is difficult to explain reasonably except by monophyletic ancestry of these birds. The lice of cuckoos are not related to those of their hosts, but possibly to those of hawks and owls and ornithologists may well not agree with lumping these together. In spite of such problems there cannot be any doubt that coevolution has contributed much to the specificity and diversity of these parasites, but even here one finds guilds. Five species of one genus of lice occur on the sanderling (*Calidris alba*) and six of the many louse species on the rook (*Corvus frugilegus*) are in one genus.

Butterflies and their foodplants have also been discussed in this connection (Ehrlich and Raven, 1964). Their coevolution can proceed to surprising lengths and a good example is the well known association between danaid butterflies and milkweeds (*Asclepias*). The larvae store cardiac glycosides which are a protective toxin produced by the plant and acquire protection themselves (Brower, 1969). Larvae reared on plants without such toxins are eaten readily by birds, but with toxins the birds rapidly learn not to touch them. The protection is carried on to the adult butterfly with similar results.

As with most evolutionary discussions we are unable to conclude anything about the actual historical origins of specialization – unless we invent a time machine to go back and look at them. We can, however, investigate selection acting now and two types of selection are relevant: anything which produces divergence between species and also selection in favour of the specialist over the 'Jack-of-all-trades'. The fourth suggestion for the evolution of specialization is special type of selection for divergence. Brower (1958) suggested that predators which learn a searching image might act in this way because it would be more difficult for such a predator to transfer his attention from one species of prey to another if they appeared different. Selection for difference as a result of predator confusion was investigated by Clarke (1962) using two species of the polymorphic snail *Cepaea* and some evidence for different polymorph frequencies was obtained where they coexisted.

There has been much discussion (Emlen, 1966; MacArthur and Pianka, 1966; Schoener, 1974) of the idea that food-limited species should be generalists because specialization would reduce their numbers and therefore natural selection could not act to produce specialization. If this were true, specialists are species whose numbers are kept low by factors other than competition and the idea that avoidance of competitors is an important cause of specialization must be discounted. But it is a gross oversimplification to say that any species is food limited. The effects of other individuals feeding may be detected at deceptively low densities. Way and Banks (1967), for example, found that fecundity and survival of the bean aphid *Aphis fabae* were reduced when numbers were far below the maximum which can live on a bean plant (Table 2.1).

Table 2.1 The numbers of bean aphids, *Aphis fabae*, developing from different numbers of apterous females at the start of the experiment (From Way and Banks, 1967)

Initial no. of females	No. of aphids at first count		No. of aphids at second count	
	Total	Total/Initial no.	Total	Total/Initial no.
2	58	29	1164	582
4	162	41	3539	885
8	370	46	5451	681
16	583	36	6349	397
32	862	27	5938	186

Generally speaking, exploitation and interference do not suddenly take effect when a critical maximum density is reached, but are progressively more severe as numbers increase. Under these circumstances, competition could stabilize population size at any level within a very wide range of density and the actual density reached would depend on the severity of other factors including density-independent ones. In other words density must be low enough for the population to replace the loss from the other factors in spite of competition. Williamson (1972) provides a thorough analysis of this point. 'Limitation' has little meaning and natural selection can be caused by any factor which alters mortality or fecundity.

Many authors have argued (Darwin, 1859; Lack, 1944; Crombie, 1947; Cain, 1954; Brown and Wilson, 1956) that competing generalists will tend to evolve into divergent specialists as the result of interaction. This is supported by the theoretical argument of Lawlor and Smith (1976). Increasing the efficiency of exploitation by substituting two specialists for two generalists using the same total range of resources is the key advantage. Divergence into new resources may also take place if these are not equally well occupied.

The degree of overlap of food requirements between coexisting species appears to be anything from nil to 100%. An example of the former occurs in *Drosophila* where *D. pachea* feeds on the Senita cactus which is toxic to other *Drosophila*. *D. pachea* needs a sterol produced by this foodplant for survival (Heed and Kircher, 1965) so there is no overlap at all with related species. Most examples of complete overlap of food are debatable on the grounds that fuller investigation would reveal differences of preference although the range of food was qualitatively the same. This is certainly true for animals, but photosynthetic plants do require the same nutrients and light. The methods of

Table 2.2 The range of food resource types and some of the consequences of specializing

Basic property	Live	Dead or non-living
Examples	Plant food, animal prey or host	Light, water, dead animals or plants, faeces
Important properties	Reactive. Exploitation affects availability of future generations	Renewal not related to exploitation
	Coevolves by interaction with exploiter	No coevolution
Basic property	Reliable →	Unpredictable or infrequent
Examples	Light, water or air-carried particles, plankton	Relatively large prey or hosts
	Leaf litter	Dead tree-trunks
	Detritus	Large carrion
	Large woody plants	Ruderals
Technique of exploitation	Static	Very mobile or very numerous dispersal phase
	Photosynthesis	
	Traps and snares	Searching behaviour
	Filters	Storing behaviour

acquiring nitrogen may differ in some, but most cannot avoid competitors by using different food.

The nature and properties of resources are worthy of further discussion and Table 2.2 summarizes the variety of energy and nutrient sources with an indication of the effects on their exploiters. Food is normally a primary resource in that it is acquired for its own sake and not as a means of gaining something else (courtship feeding is the main exception), but space can be primary or secondary. A special breeding site such as a nesting hole for a bird or a special protective site could be regarded as primary resources, but a territory is frequently an area defended in order to monopolize the food it contains. Time is sometimes also called a resource, but is always secondary. It is important in this context only as time in which to feed or search for food and space. These are not the only differences between resources and another which might prove interesting is that space is not consumed in the way food is, but is available to another individual when vacated.

One point is worth re-emphasizing to end this section. Food, space, time and habitat type are not independent of each other and caution is needed in estimating resource overlap between species.

2.2   Stratification

The subdivision of major habitats into zones containing different species is brought about by several causes which may act together. (1) Tolerance of, or adaptation to, only part of the range of conditions available. (2) Habitat selection by the animals. (3) Restriction of the range of one species by another as the result of interference, or possibly other interactions outside the scope of this discussion, such as predation. The zonation of the seashore is perhaps the best known and most striking example of subdivision of a habitat with a continuous gradient. The barnacles investigated by Connell (1961) are zoned as the result of two of these processes. *Balanus balanoides* is absent from the upper shore because of intolerance of longer exposure and *Chthalamus stellatus* is excluded from the lower shore by interference from *B. balanoides*. Neither habitat selection nor exploitation of the same food are directly relevant to this example and the resource shared is space in which to live.

In freshwater the planarians may occupy zones of streams and a clear example was provided by Beauchamp and Ullyott (1932) working in the Balkans on *Planaria montenegrina* and *P. gonocephala*. The former species occurred from the springs to a level where the temperature was 16–17°C when alone, but only to 13–14°C when *P. gonocephala* was present in the same stream. *P. gonocephala*, on the other hand, occurred from the springs to 21–23°C when alone, but only from a narrow zone of overlap at 13–14°C to 21–23°C when together in the same stream with *P. montenegrina*. Adaptation to specific temperature ranges was important, but not the only cause of the zonation and either the species interact to exclude each other or some unknown factor causing the absence of one species also affects the range of the other.

The gradients produced by plant succession can also be partitioned by animal species which are not specialist herbivores. Lack (1933) working on passerine birds of conifer plantations showed that the changes with age of plantation were rapid (Table 2.3) and proposed that habitat selection played

Table 2.3   The approximate percentage of bird population formed by these species in grassland and conifer plantations of different age in Breckland, England (From Lack, 1933)

	Wheatear	Stock dove	Lapwing	Stone curlew	Skylark	Meadow pipit	Whinchat	Willow warbler	Whitethroat	Wren	Dunnock	Blackbird	Chaffinch	Goldcrest
Short turf	40	5	5	15	30	—	—	—	—	—	—	—	—	—
Thin grass	5	<3	5	5	65	20	—	—	—	—	—	—	—	—
4 and 5 yr-old plantation	<3	—	—	<3	40	30	20	15	—	—	—	—	—	—
6-yr-old	—	—	—	<3	25	15	10	30	5	5	—	—	—	—
7-yr-old	—	—	—	—	10	5	5	45	10	5	5	5	—	—
8-yr-old	—	—	—	—	—	5	—	50	20	10	5	—	5	<3
9-yr-old	—	—	—	—	—	<3	—	65	5	5	10	10	—	5

the major part. In many insect groups the shading effect of taller plants and the consequent reduction in soil surface temperature appears to be the most important change. In ants of genus *Lasius* the succession has become obligatory. Founding queens of *L. niger* choose shadeless sites to start colonies and are dependent on insolation for brood rearing (Pontin, 1960). This species behaves as a ruderal colonizing disturbed areas and is followed by *L. umbratus* and *L. fuliginosus* which are shade tolerant because they use metabolic heat to maintain a higher than ambient nest temperature. *L. umbratus* queens found their colonies by temporary social parasitism of *L. niger* and cannot found colonies without the assistance of workers (*see* Donisthorpe, 1927). *L. fuliginosus* queens parasitize *L. umbratus* in turn and need aphid populations on woody plants for success with their foraging technique using scent trails above ground.

The plants themselves are commonly subdivided by their fauna as was described by Hartley (1953) working on foraging tits (*Parus*) (Fig. 2.2). The two most abundant species *P. major* and *P. caeruleus* feed for a higher proportion of their time in the lower layers of woodland and in the canopy respectively. Snow (1949) suggested that the blue tit can hang on smaller twigs than can be used by the heavier great tit, automatically producing differences in foraging strategy and possibly offsetting any advantage of large size.

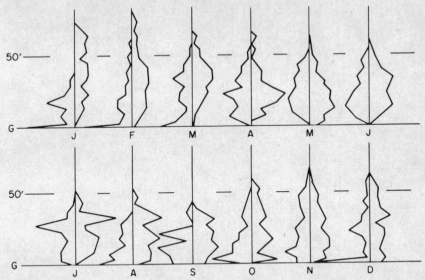

Figure 2.2   The relative frequency of feeding at different heights in woodland by great tit (left) and blue tit (right) during each month of the year. G, ground level. (*From* Hartley, 1953.)

Such subdivision is widespread among insects. Broadhead and Wapshere (1966) working on two very similar Psocoptera found that they tended to oviposit in different zones of trees. *Mesopsocus unipunctatus* laid the majority of its egg batches on thinner twigs near the extremities of larch (*Larix*) branches while *M. immunis* oviposited relatively infrequently on the same twigs, but used much thicker twigs. They coexisted on the same trees in some areas, but the egg distribution of each species was unchanged in the absence of the other. The difference between them is not the result of any direct interaction going on at present and may be the result of natural selection acting in the past to increase specialization.

Askew (1962) working on the distribution of spangle galls on oak leaves found another example of zonation produced by oviposition behaviour. Three species of *Neuroterus*, *N. numismalis*, *N. laeviusculus* and *N. lenticularis*, showed a positive correlation of occurrence on different trees, but some partitioning of the parts of the trees. *N. numismalis* was more abundant near the tops of young trees, more abundant near the periphery of the foliage and markedly preponderant near the tips of the leaves. The differences of distribution on leaves were not dependent on the presence of other species (Fig. 2.3), but there was a suggestion of a shift when they coexisted. Space may be restricting on some leaves, but the relative importance of food and space is unknown.

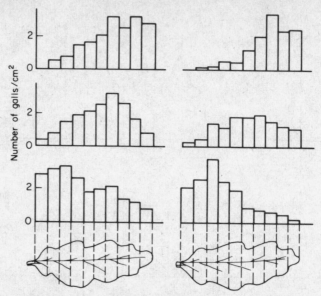

Figure 2.3 Density (per cm²) of spangle galls of three *Neuroterus* spp. along oak leaves when alone (left) or with congeners on the same leaf (right). Top, *N. numismalis*; centre, *N. lenticularis*; bottom, *N. laeviusculus*. (*From* Askew, 1962.)

In some cases animals are clearly restricted to part of a resource gradient by inherited morphological differences of their own. A particularly striking example is shown by ovipositor length of the ichneumonids of genus *Megarhyssa* (Townes and Townes, 1960; Heatwole and Davis, 1965). These parasitize insect larvae in dead logs and insert their ovipositors full length when egg laying in hosts. The succession of fungal attack and host availability starts at the outside and progresses towards the interior thus producing a succession of *Megarhyssa* species with ovipositor lengths of around 40, 75 and 115 mm, respectively. Tongue length of bees may also restrict the range of flowers available to different bee species and produce some resource partitioning (e.g. Heinrich, 1976) (Fig. 2.4). The bumblebees with short tongues tend to be restricted to open flowers and the long-tongued bees show some specialization on tubular flowers. Learning plays some part in their specialization and they concentrate on the flowers with most nectar. The specialization of individual bees is not complete and this gives them the flexibility to change to more rewarding flowers as they come into season or switch to flowers with honey flow restricted to part of the day.

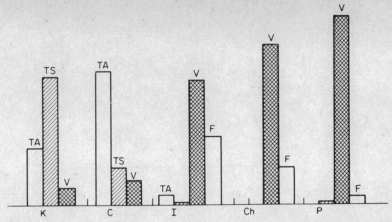

Figure 2.4 Relative frequency of four bumblebee species visiting five different flower types. In *Bombus terricola* (TA) the tongue is 6.5 mm long; in *B. ternarius* (TS) 6.2 mm; *B. vagans* (V), 8.0 mm; and *B. fervidus* (F) 10.8 mm. *Kalmia* (K) is an open flower, *Cephalanthus* (C) has a 7-mm floral tube, *Impatiens* (I) a 20-mm tube, *Chelone* (Ch) a 25-mm tube, and *Pantederia* (P) a 5-mm tube. Tube length is clearly important to the bees, but *Pantederia* shows that other factors are also important. (*From* Heinrich, 1976.)

It is more usual for environmental discontinuities to determine the bound-aries between the activities of the two species. A obvious example is the soil surface. The common shrew (*Sorex araneus*) and pigmy shrew (*S. minutus*) occur in the same woodland in England, but *S. araneus* is much more subterranean while *S. minutus* has territories in the superficial litter overlap-ping those of *S. araneus* (Michielsen, 1966). The difference is reflected in their food (Pernetta, 1976), but they do have some food in common and it is not known whether they affect each others' population density or distribu-tion. The same applies to most of the behaviour examples quoted here, but the ants *Lasius flavus* and *L. niger* show a parallel division of their habitat and they do interact ecologically (Chapter 4). Stratification in time can be based either on a seasonal difference or a diurnal difference in activity. The first example of specialization cited – the chalk hill blue butterfly – has a congener the adonis blue *L. bellargus* coexisting on the same food plant in many of its more southerly British localities. They differ seasonally because *L. bellargus* is bivoltine with adult flight periods before and after *L. corydon*. There are many similar examples of seasonal segregation and it should not be assumed that any of them are ecologically isolating.

Ants may sometimes show differences in foraging rhythm and the two dolichoderines observed by Hunt (1974) in Chile illustrate this (Fig. 2.5). *Tapinoma antarcticum* forages at high temperature in the central period of each day while *Dorymyrmex antarcticus* forages mainly just after sunrise and

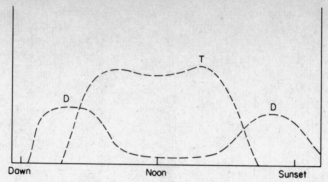

Figure 2.5 Foraging activity of two Chilean ant species, *Tapinoma antarcticum* (T) and *Dorymyrmex antarcticus* (D) from dawn to sunset. (*From* Hunt, 1974.)

in the evening at medium temperatures. Artificial shading of the nests reduced *Tapinoma* foraging, but prolonged the period of *Dorymyrmex* foraging showing that the proximate factor was insolation. *Tapinoma* recruits foragers to a feeding site rapidly and is aggressively dominant to other species; this may offset the limitations of its activity rhythm.

References

Abbott, I., Abbott, L. K. and Grant, P. R. (1977). Comparative ecology of Galapagos ground finches *Geospiza*. Evaluation of the importance of floristic diversity and interspecific competition. *Ecol. Monogr.* **47**, 151–84, (The Ecological Society of America).

Askew, R. R. (1962). The distribution of galls of *Neuroterus* (Hym: Cynipidae) on oak. *J. Anim. Ecol.* **31**, 439–55.

Beauchamp, R. S. A. and Ullyott, P. (1932). Competitive relationships between certain species of fresh-water triclads. *J. Ecol.* **20**, 200–8.

Broadhead, E. and Wapshere, A. J. (1966). *Mesopsocus* populations on larch in England – the distribution and dynamics of two closely-related coexisting species of psocoptera sharing the same food resource. *Ecol. Monogr.* **36**, 327–88.

Brower, L. P. (1958). Bird predation and foodplant specificity in closely related procryptic insects. *Amer. Naturalist* **92**, 183–7.

Brower, L. P. (1969). Ecological chemistry. *Scient. Amer.* **220**, 22–9.

Brown, W. L. and Wilson, E. O. (1956). Character displacement. *Syst. Zool.* **5**, 49–64.

Cain, A. J. (1954). *Animal Species and their Evolution*. London: Hutchinson.

Clarke, B. C. (1962). Balanced polymorphism and the diversity of sympatric species. *Systematics Association Publication* **4**, 47–70.

Connell, J. H. (1961). The influence of interspecific competition and other factors on the distribution of the barnacle *Chthalamus stellatus*. *Ecology* **42**, 710–23.

Crombie, A. C. (1947). Interspecific competition. *J. Anim. Ecol.* **16**, 44–73.

Darwin, C. (1859). *The Origin of Species by Means of Natural Selection*. Harvard Facsimile 1st edition, 1964.

Donisthorpe, H. St J. K. (1927). *British Ants their Life-History and Classification*. London: Routledge.

Ehrlich, P. R. and Raven, P. H. (1964). Butterflies and plants: a study in coevolution. *Evolution* 18, 586–608.

Emlen, J. M. (1966). The role of time and energy in food preference. *Amer. Naturalist* 100, 611–17.

Ford, E. B. (1945). *Butterflies*. London: Collins.

Hartley, P. H. T. (1953). An ecological study of the feeding habits of the English titmice. *J. Anim. Ecol.* 22, 261–88.

Heatwole, H. and Davis, D. M. (1965). Ecology of three sympatric species of parasitic insects of the genus *Megarhyssa* (Hymenoptera: Ichneumonidae). *Ecology* 46, 140–50.

Heed, W. B. and Kircher, H. W. (1965). Unique sterol in the ecology and nutrition of *Drosophila pachea*. *Science* 149, 758–61.

Heinrich, B. (1976). The foraging specialisations of individual bumble bees. *Ecol. Monogr.* 46, 105–28, (The Ecological Society of America).

Hunt, J. H. (1974). Temporal activity patterns in two competing ant species (Hymenoptera: Formicidae). *Psyche J. Entomol.* 81, 237–42.

Janzen, D. H. (1977). The interaction of seed predators and seed chemistry. In: *Labyrie, V. Comportement des insectes et milieu trophique*. *Coll. Int. C.N.R.S.* 265, 415–27.

Lack, D. (1933). Habitat selection in birds with special reference to the effects of afforestation on the Breckland avifauna. *J. Anim. Ecol.* 2, 239–62.

Lack, D. (1944). Ecological aspects of species-formation in passerine birds. *Ibis* 86, 260–86.

Lack, D. (1947). *Darwin's Finches*. Cambridge University Press.

Lawlor, R. L. and Smith, J. Maynard (1976). The coevolution and stability of competing species. *Amer. Naturalist* 110, 79–99.

MacArthur, R. H. and Pianka, E. (1966). An optimal use of a patchy environment. *Amer. Naturalist* 100, 603–9.

Michielsen, N. C. (1966). Intraspecific competition in the shrews *Sorex araneus* L. and *S. minutus* L. *Archs Néerl. Zool.* 17, 73–174.

Pernetta, J. C. (1976). Diets of the shrews *Sorex araneus* and *Sorex minutus* in Wytham grassland England. *J. Anim. Ecol.* 45, 899–912.

Pontin, A. J. (1960). Field experiments on colony foundation by *Lasius niger* (L.) and *L. flavus* (F.) (Hymenoptera: Formicidae). *Insectes Sociaux* 7, 227–30.

Root, R. B. (1967). The niche exploitation pattern of the blue-gray gnatcatcher. *Ecol. Monogr.* 37, 317–50.

Ross, H. H. (1957). Principles of natural coexistence indicated by leafhopper populations. *Evolution* 11, 113–29.

Rothschild, M. and Clay, T. (1952). *Fleas, Flukes and Cuckoos*. London: Collins.

Schoener, T. W. (1974). Competition and the form of habitat shift. *Theor. Popul. Biol.* 6, 265–307.

Snow, D. W. (1949). Jämforände studier över våra mesarters näringssökande. *Vår Fågelv.* 8, 156–69.

Townes, H. and Townes, M. (1960). Ichneumon-flies of America north of Mexico: 2. *Bull. U.S. Nat. Mus. Washington.*

Way, M. J. and Banks, C. J. (1967). Intra-specific mechanisms in relation to the natural regulation of numbers of *Aphis fabae* Scop. *Ann. appl. Biol.* **59**, 189–205.

Williamson, M. H. (1972). *The Analysis of Biological Populations.* London: Arnold.

3 The causes of competition: interference

Where the shared resources are indivisible such as space in which to live or where one individual prevents another from taking a share, the outcome is decided by direct aggression or simple avoidance of a dominant individual by a subordinate one. Nicholson (1955) used the terms 'scramble' and 'contest', when discussing intraspecific competition, for the two extremes of this behaviour. A scramble occurs when the resources are shared out equally so that all get insufficient if the resources run out. A contest occurs when a proportion of the individuals always get enough while the others go short and may die. Where individual size and fecundity are flexible, intermediates occur (Pontin, 1961) and there may well exist a continuous spectrum of examples between these extremes. Exploitation alone produces a scramble and interference converts this, more or less, to a contest.

Individuals of one species interfere with individuals of another species by using a wide variety of behaviour which may be listed as follows:
(1) Simple overgrowth of a larger or quicker growing sessile species over a smaller. (2) Directly physical attack. (3) Chemical toxins or deterrents which may be the same as their protection against predation. (4) Ritualized display using visual or auditory stimuli similar to their intraspecific spacing behaviour. Whichever method, or combination of methods, is used the results are mosaic distributions or sharpened zonal boundaries between species. Many examples have been described and it is convenient to discuss them under three headings: stationary animals and plants, territorial animals and, thirdly, fully mobile animals.

3.1 Stationary organisms

The causes of plant competition can be listed as: overshadowing, depletion of minerals and water, deposition of leaf litter and release of toxins (allelopathy). Grime has recently (1979) reviewed these factors in the more general context of survival strategies and, in the case of bracken (*Pteridium aquilinum*), provides an excellent illustration of them acting together. By using storage of the previous year's production its tall fronds can quickly overshadow other herbs and can delay succession because it takes more than one year for woody plants to grow large enough supporting structures (Al–Mufti *et al.*, 1977). It deposits a thick litter which further inhibits other plants, but bracken has precisely the growth form to counteract its own litter. It grows stout shoots very quickly. Large areas of continuous cover are produced and the term 'dominance' is used to describe such monopolization. Root competition between trees has long been known (e.g. Fricke, 1904) and trenching around trees to cut roots can increase the growth of young neighbours.

Depletion of nutrients will of course inhibit intraspecifically as well as interspecifically and toxin production may also rebound on the very plants producing inhibitors. In both cases therefore field evidence is needed before concluding these factors are causes of dominance. It is very difficult to distinguish them and, although many cases of toxin production have been demonstrated in laboratory cultures, field evidence of toxic effects is inconclusive (*see* Harper, 1977). The Colorado desert shrubs worked on by Went (1942) illustrate the problem. Some species including *Franseria dumosa* and *Laurea tridentata* normally have many annuals growing under their cover, but *Encelia farinosa* has very few. *Encelia* produces an identifiable toxin, but *Franseria* when tested was shown to be even more toxic. The wind-blown debris which collects around the basal branches of *Franseria* apparently provides a better place for germination of annuals and further field experimentation is needed. Toxins could be very important in deciding the winner in plant competition where stability from differences in exploitation is much less likely than with animals. The antibiotics produced by moulds and bacteria have become household words and need no further discussion here, but presumably autotoxicity is negligible to allow ecological dominance.

The sessile animals of the seashore show some parallels with higher plants. Overgrowth and toxin production have both been reported and the work of

Table 3.1   Effects of homogenates of sponges and ascidians on ectoprocts in aquaria

	Ectoproct species tested		
Homogenates Sponges	*Steganoporella magnilibris*	*Stylopoma spongites*	*Reptadeonella violacea*
Mycale laevis	+ +	0	
Tenaciella?	0	0	
Toxemna?	0	+ +	
Halisarca?	0	0	
Plakortis sp.	+ +	0	+ +
Agelas sp.	+	0	
Ectoplasia ferox	0	+ +	+ +
Colonial ascidians			
Didemnum sp.	0	0	
Ascidian '2'	+	+ +	

0, no apparent effect; +, loss of movement; + +, deterioration of polyps. The other combinations were not tried. (Selected from Jackson and Buss, 1975.)

Connell (1961) (Chapter 5) provides an excellent example of a faster growing species (*Balanus balanoides*) smothering or displacing a smaller one (*Chthalamus stellatus*). Gordon (1972) describes overgrowth of ectoprocts by ascidians and sponges.

Where overgrowth is not possible allelopathy is potentially important. Jackson and Buss (1975) describe a coral reef community where five out of nine sponge species and one of the two ascidian species produce chemicals which cause ectoproct mortality in laboratory tests of homogenates (Table 3.1). The results appear to show some specificity and none of the solitary animals tested showed any ill effects from the homogenates. There is no reason to suppose that it will be any easier to provide conclusive evidence for animal allelopathy in the field than it has been for plants.

Menge (1976) describes an interesting example of interference by whiplash effect of fucoids reducing *Balanus* recruitment. This has no obvious value to the seaweeds and there may well be many examples of accidental interference by mechanical damage in other habitats, for example trampling effects of large grassland herbivores. Shore animals are able to be sessile because food is continually brought to them and this makes food overlap less important as a primary cause of competition, but it would be wrong to assume that interference is acting alone. Space becomes the important resource, but

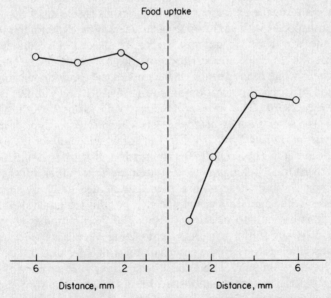

Figure 3.1 Uptake of food by neighbouring colonies of ectoprocts at different distances (mm) from their meeting margins. *Onychocella alula* (left) is little affected by closeness of *Antopora tincta* (right), but *A. tincta* is deprived of food particles. Ordinate is radioactivity following intake of labelled particles. (*From* Buss, 1979, reproduced here by courtesy of *Nature*.)

overgrowth also enables the lower animal to be put into a 'food shadow' analogous to one plant taking the light from another. The feeding advantage then enables more differential growth and instability by positive feedback follows. Buss (1979) describes the method used by *Onychocella alula* to dominate another, smaller, ectoproct *Antropora tincta*. Their lophophores extend only a fraction of a millimetre above the surface, but their feeding currents can dominate the flow around them. The larger ectoproct showed no reduction in acquisition of radioactive food particles next to the small one (Fig. 3.1), but *Antropora* showed a definite reduction in clearance rate when in contact with *Onychocella*. This was confirmed by observing milk in the feeding currents and a band of several millimetres over the small ectoproct margin was cleared by the large one.

3.2   Animals with interspecific territories

These are special cases which are in some ways intermediate between sessile and fully mobile species because they are mobile around a fixed nest site. They have corresponding advantages over mobile species for ease of study and field experimentation. The main examples are found in ants and birds, but there are other cases where aggression is shown to other species in a territorial pattern, for example, male dragonflies (Moore, 1964).

Those ant species with very populous nests dominate their habitats in many parts of the world and frequently several species are codominant. Pickles (1944) describes interspecific territorial battles between *Messor barbarus* and *M. aegypticus* while many authors have studied the three-dimensional mosaic distribution of tropical plantation ants (e.g. Way, 1953; Vanderplank, 1960; Majer, 1972, 1976). The ants of Ghana cocoa farms include *Oecophylla longioda*, *Macromischoides aculeatus*, six *Crematogaster* spp., *Camponotus acvapimensis* and *Platythyrea frontalis*, any of which could locally exclude the other codominants (Majer, 1972). Aggression is shown at the boundaries between territories belonging to different nests whether of the same or different species and these boundaries may change daily. A succession of some species with vegetation growth occurs, but this is a much slower cause of changes. It is difficult to be sure whether the distribution is a true mosaic determined by the initial pattern of colonizing queens or whether there is some undescribed pattern in the environment which decides the species which becomes locally dominant. There must be considerable inertia prolonging the colonizing pattern because established nests kill any queens of any species which arrive later (*see* Pontin, 1960). Laboratory studies of ant aggression have been made by de Vroey (1979) and Moxon (1980) and the latter shows that aggression by *Myrmica scabrinodis*, compared with other *Myrmica* species, has clearly been evolved to cope with *Lasius flavus*, its normal neighbour in the field.

As well as direct physical attack, ants typically use chemical repellents (*see* Wilson, 1971) against other ants. This technique can be very effective in allowing small species such as *Tapinoma erraticum* to drive off much larger ants from feeding sites (*see* Donisthorpe, 1927). Lofqvist (1977) describes the action of the chemical defence and offence system in *Formica sanguinea* and *F. rufa*. Formic acid is the major constituent in volume, but wetting agents are important additives for it to enter tracheal systems and kill ants. *F. rufa* chemicals are more toxic to both species than those of *F. sanguinea* and *F. sanguinea* seems more sensitive to the chemicals of either. *F. sanguinea* is clearly superior in direct conflict in spite of this and appears much more active, better coordinated in attack and often larger (although Lofqvist found that it was smaller). Its normal mode of life includes regular raiding of other *Formica* nests with the capture of pupae for 'slaves' so it may well be better adapted for inter-ant conflict. In contrast, some small ant species, for example *Leptothorax*, are frequently observed to nest within the territories of much larger *Formica* or *Lasius* which do not react to them. The *Leptothorax* may crouch in a crevice, but perhaps they are chemically 'invisible' and survive by avoiding conflict.

The large owl limpet *Lottia gigantea* defends a very definite territory which is its feeding area (Stimson, 1970). Other grazing limpets such as *Acmaea* as well as predatory gastropods, anemones and barnacles are pushed off an area of 1000 cm² and a thick film of algal food cropped by *Lottia* grows in the absence of other grazers.

Some birds show normal intraspecific territorial behaviour towards members of other species (Simmons, 1951) and this may result in a mosaic of territories where interspecific exclusion is as effective as intraspecific exclusion. The best known examples are the meadowlarks *Sturnella magna* and *S. neglecta* (Lanyon, 1956) which have a narrow zone of sympatry and show interspecific territory in spite of differing song and also the woodpeckers *Centurus carolinus* and *C. aurifrons* with more similar songs and mutually exclusive territories in a small area of overlap (Selander and Giller, 1959).

Two interpretations of this phenomenon have been made. Murray (1971, 1976) suggested that the examples are accidental and result from misdirected intraspecific behaviour. He claims this to be maladaptive for one species which could evolve dominance over the other and displace it from the whole area of overlap. Cody (1969), on the other hand, suggested that adaptive convergence between similar species should occur to more effectively produce mutually exclusive territories. Interference is argued here to be basically unstabilizing and this view favours Murray, but, whether convergence occurs or not, it is an immediate advantage to exclude any individuals which exploit the same resources even if they are different species.

One cannot expect to generalize about interspecific territories because of the different functions of territory in different examples. Catchpole (1973, 1978) working on *Acrocephalus* warblers used replay of recorded song and found that interspecific response occurred only when later arrivals invaded an area occupied by a territory holder. These birds, although showing strongly held mutually exclusive nesting territories, fed outside them in a variety of habitats including those not used for nesting (Fig. 3.2). The overall density of warblers may nevertheless be reduced and breeding success of individual pairs of either species improved.

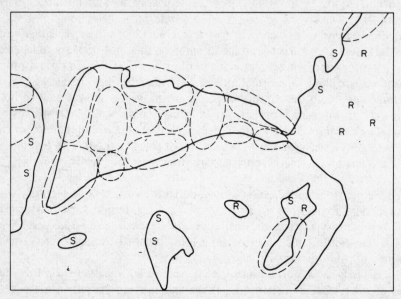

Figure 3.2 Interspecific territories of reed warbler (R) with smaller territories and sedge warbler (S) in an area of 350×250 m. Much of this area is water in disused gravel workings and the shoreline is very irregular with small islands. Some feeding stations outside the defended territories are indicated by the foragers' initial letters. (*From* Catchpole, 1973.)

Edington and Edington (1972), on the other hand, found that the less exclusive territories defended by redstart *Phoenicurus phoenicurus* and pied flycatcher *Ficedula hypoleuca* contained non-overlapping mutually exclusive feeding areas (Fig. 3.3). They also described fully exclusive interspecific territories between wood warbler *Phylloscopus sibilatrix* and willow warbler *P. trochilus* although habitat zonation plays a part in this case.

Where nest sites are special and relatively few much aggression may be observed between species. Slagsvold (1978) reports an interesting experiment where great tits *Parus major* were allowed to occupy nestboxes and the supernumerary boxes were then made unsuitable for the later nesting pied flycatcher. The tits prevented most of the flycatchers from breeding and more aggression was recorded than occurred when excess nestboxes were available.

Other examples of rather different interspecific spacing have been described. Hubbell and Johnson (1977) worked on eight species of stingless bees in Costa Rican forest. All nested in tree cavities with the larger species choosing trees with larger bole diameter, but no preference for tree species was found. There was every indication of superabundance of nest sites, but four species, *Trigona silvestriana*, *T. fuscipennis*, *T. fulviventris* and *T. pectoralis*, showed very uniform colony spacing. *T. dorsalis*, *T. capitata*,

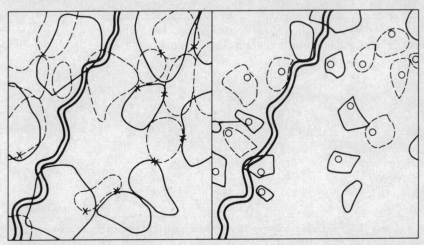

Figure 3.3 Territories (left) and feeding areas (right) of pied flycatcher (broken outlines) and redstart (continuous outlines) in a valley woodland of approx. 15 ha. Sites of interspecific fighting are indicated by crosses and nestholes by circles. (*From* Edington and Edington, 1972.)

T. testaceicornis and *T. frontalis* appear randomly distributed and this reflects a clear-cut behavioural difference between these groups. The second set are solitary foragers while the others can recruit foragers to group at rich sources by laying pheromone markers. Nest spacing is produced, in *T. fulviventris* at least, by workers using the pheromone markers to attract nestmates to a new nest site. The marks also attract workers from any nearby rival colonies and aggressive behaviour follows. The bees form opponent pairs and fly vertically up with some grappling in the air. Injury does not normally occur and the behaviour gave the impression that it had been ritualized. If a group of bees attempts to found a colony in a tree already occupied more prolonged wrestling occurs and the authors recorded a large number of dead after one such battle. New nests will be much more easy to establish away from pre-existing colonies and the nest spacing observed can be produced in this way.

Marking pheromones can be detected by different species and *T. ful-viventris* responds to the pheromones of *T. silvestriana* and *T. fuscipennis* by agitation and flight. Interspecific spacing was therefore examined. Nearest-neighbour analysis was used following the method of Clark and Evans (1954) which compares the actual average distance to nearest neighbour with that expected in a random population of the same overall density. The populations of pairs of species were lumped together and tested as a single population for comparison with nearest-neighbour indices of single species populations. *T. fulviventris* and *T. fuscipennis* still showed a high degree of spacing when combined (Table 3.2) and *T. silvestriana* paired with either of these also

Table 3.2 Nearest-neighbour indices of *Trigona* bee colonies comparing pairs of conspecific neighbours with pairs of different species

	T. silvestriana	*T. fuscipennis*	*T. fulviventris*	*T. pectoralis*	*T. dorsalis*
T. silvestriana	2.76‡				
T. fuscipennis	1.61†	1.84§			
T. fulviventris	1.34†	1.64§	1.72†		
T. pectoralis	1.13	1.26	1.31	1.61*	
T. dorsalis	1.08	1.22	1.38	0.915	0.982

High indices indicate greater spacing out of neighbours than can be expected in a random population. *P, 0.05; †P, 0.01; ‡P, 0.005; §P, 0.001. (From Hubbell and Johnston, 1977.)

showed interspecific spacing, but not the other combinations. The indices of intraspecific dispersion have the same rank order as the frequency with which they were observed to show aggression and the result is saturation of the habitat in a manner very similar to full territorial behaviour. It is perhaps surprising that so many species occur together without displacement by

Table 3.3 Minimal distances between groups of the same species compared with minimal distances between groups of different species (From Krämer, 1973, reproduced here by courtesy of *The Journal of Wildlife Management*.)

	No. of observations		
Species pair	>50 yd (45.7 m)	25–50 yd (22.9–45.7 m)	<25 yd (22.9 m)
Mule deer/mule deer	324	234	0
White-tailed deer/white-tailed deer	166	85	0
Mule deer/white-tailed deer	88	32	27
Mule deer/any species except deer	99	19	11
White-tailed deer/any sp. except deer	46	7	3
Deer/Coyotes	38	4	1

$P<0.01$ of Mule deer having same intra- and interspecific spacing.

dominance and that the non-spacing species have managed to remain as though unnoticed.

Grazing mammals may show spacing behaviour even less like interspecific territories. Krämer (1973) estimated the distances between homospecific groups of white-tailed deer, *Odocoileus hemionus*, and mule deer, *Odocoileus virginianus*, and compared these with interspecific distances (Table 3.3). The interspecific spacing is clearly less well marked than intraspecies spacing and coexistence of the species is to be expected.

3.3 Fully mobile species

Priority of access to resources is the usual result of interference between individuals of mobile species in contrast to the local monopoly of resources established by sessile or territorial species where space is an important resource. Morse (1974) calls this priority 'social dominance' instead of simple 'dominance' which is commonly applied to sessile organisms establishing pure stands of one species. He gives an interesting table of examples selected because they show some evidence of the effect of social dominance on exploitation, but it is often not easy to state which species is consistently dominant to another. Individual variation in age, size and motivation or changes in environmental factors may reverse the direction of social dominance. There are many anecdotal examples, but only a few of the better-documented cases will be cited here to give an outline of the enormous variety of interaction involved.

Williams (1975) and Purcell (1977) present evidence for the function of fighting tentacles in sea anemones. They are not used in feeding and Purcell's suggestion that they should be called fighting tentacles instead of 'catch' tentacles is adopted here. *Metridium senile* develops the tentacles nearest the mouth in response to genetically different anemones, not clonemates, and the feeding nematocysts are replaced by atrich and holotrich nematocysts. Attacked anemones are severely injured. Genera *Haliplanella*, *Diadumene*, *Cereus* and *Sagartia* are also reported to develop fighting tentacles and they are used interspecifically as well as between members of different clones of one species. Another interesting seashore example concerns the hermit crabs *Clibinarius vittatus* and *Pagurus* spp. of Texas (Wright, 1973). *Clibinarius* is dominant when fighting with *Pagurus* for a shell unless the shell bears a hydroid colony (*Hydractinia* or *Podocoryne*) which repels *Clibinarius* especially after previous experience of being stung. *Pagurus* are not stung by hydroids and frequently (20–30%) have them on their shells, thus reversing dominance. *Clibinarius* very rarely has hydroids on its shell and probably exposes itself too much at low tide for hydroid survival.

Interference, like exploitation, may occur between taxonomically widely separated species. Primack and Howe (1975) describe interference between pollinating skipper butterflies and a hummingbird *Amazilia tzacatl* at a *Stachytanpheta* hedge. *Amazilia* chased skippers which flew lower or left the hedge and removal of the *Amazilia* resulted in higher than usual foraging by the skippers (Fig. 3.4).

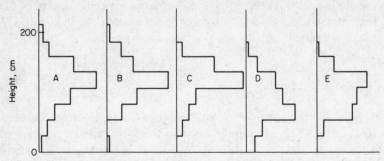

Figure 3.4 Height distribution of flowers (A), hummingbird foraging (B), chases of skipper butterflies by hummingbirds (C), skipper foraging with hummingbirds present (D), and skipper foraging with hummingbirds removed (E). (*From* Primack and Howe, 1975.) (Reproduced by permission of the Association for Tropical Biology.)

Dominance of bats over nighthawks *Chordeiles minor* when insect hunting in a cone of spotlight illumination was observed by Shields and Bildstein (1979). 75% of the bat aggressions and 94% of the nighthawk aggressions were against nighthawks and some displacement of nighthawks from the richer lower part of the light cone was measured when the bats were present.

Mammal studies are usually behavioural and lack assessment of the ecological effects of interference because of the difficulty of population measurement. The forest-living *Clethrionomys gapperi* and grassland *Microtus pennsylvanicus* have received much attention because of occasional invasion of each other's habitat (Grant, 1972; Iverson and Turner, 1972; Turner, Perrin and Iverson, 1975). Temporary coexistence in winter ceases when intraspecific aggression reaches its peak in spring at reproductive maturity. *Microtus* may become dominant over *Clethrionomys* in laboratory-staged fights at this time and aggression increased four-fold compared with winter condition. Resources partitioning is less important than the unstabilizing effects of interference in this case.

Laboratory mice are killed by *Rattus norvegicus* (Karli, 1956) and this response is increased by food deprivation (Milner, 1976), but this behaviour is of unknown significance in the wild. It could represent an interference mechanism evolved to eliminate a species with overlap in food exploitation or perhaps it is predatory behaviour. The large size difference determines dominance, but mice may be able to avoid rats in a more complex environment.

A very different example illustrates evolution of special morphological adaptations which eliminate other resource sharing individuals. Salt (1961) and Fisher (1961) describe examples of elimination of parasitoid insect larvae when in the same host with other individuals. Very large mandibles are possessed in the first instar of some species (Fig. 3.5) and the first of these to hatch attacks the others ensuring that only one parasite per host survives.

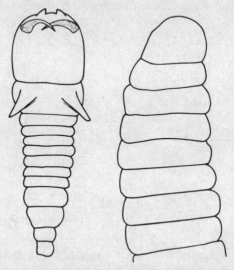

Figure 3.5   First instar (left) and second instar (right) larvae of the parasitoid *Opius fletcheri* showing fighting jaws which are lost when no longer required. (*From* Willard, 1920.)

Occasionally, different techniques are used in intraspecific aggression from those used interspecifically. Giraffe use small horns against other giraffe, but kick predators (Dagg and Foster, 1976) and this distinction appears to be a feature of those species which evolve very dangerous interspecific techniques. If this is found in interspecific competitive interactions it should improve dominance.

3.4   The intervention of 'third party' organisms

There are several very interesting cases where non-food species produce interactions which simulate competition or affect its outcome. Grant and Grant (1965) argued that competition between plant species for pollinators was important and Levin and Anderson (1970) suggested that sequential flowering of coexisting species might be evolved as a consequence. When one species ceases flowering there will then be a population of pollinators ready to exploit the next species so that there is a disadvantage in having a gap between their flowering periods (Heinrich and Raven, 1972). There should therefore be a balance between selection for separation and selection for mutualism. Waser and Real (1979) produced some results (Table 3.4) which showed reduced seed setting by *Ipomopsis* when the earlier flowering *Delphinium* was at a lower density. Reduced pollinating visits by the

Table 3.4   The relationship between number of *Impomopsis* seed set, the visits of polinators and the density of earlier flowers (From Waser and Real, 1979, reproduced here by courtesy of *Nature*.)

	Peak density of flowers per 4 m² in 6 plots		No. of Ipomosis seeds per flower	Mean no. of Hummingbird visits per 10-min census Broad tail		Rufous
	Delphinium	Ipomopsis		Overall	Late	
1975	25.0	29.3	10.1	10.2	8.6	7.1
1976	30.8	27.8	12.4			
1977	0.8	18.7	7.6	8.0	6.9	7.2
1978	7.2	39.3	7.1	5.0	5.0	5.0

Regression analysis of seed number on *Delphinium* density $P<0.001$ and of seed number on broad tail visit number $P<0.01$.

broad-tailed hummingbird were also correlated with low density of *Delphinium* and this lower activity persisted into the flowering period of *Ipomopsis*. The rufous hummingbird which is absent for the *Delphinium* flowering period showed similar activity in each year thus strengthening the evidence that *Delphinium* density affects the number of pollinating visits to *Ipomopsis* by broad-tails.

A completely different example was described by Springett (1968). Phoretic mites *Poecilochirus* are normally found on carrion beetles of genus

Table 3.5   The results of infesting standard mouse (*Apodemus*) corpses with different combinations of carrion beetles, blowfly eggs and *Poecilochirus* mites (From Springett, 1968).

	Cultures with successful Necrophorus	Cultures with successful mites	Cultures with successful blowflies
30 mites, 100 blowfly eggs		0	0
30 mites, 100 blowfly eggs, pair of Necrophorus	6	8	0
30 mites, pair of Necrophorus	7	8	–
100 blowfly eggs	–	–	8
Pair of Necrophorus	8	–	–
100 blowfly eggs, pair of Necrophorus	0	–	8

Necrophorus and these are predators of the carrion beetles' main competitors – they eat blowfly eggs. Laboratory cultures of a pair of beetles with blowfly eggs on a mouse carcass were unsuccessful even though the beetles attempted to eat fly larvae, but similar cultures with mites were successful in allowing beetle larvae to mature (Table 3.5). Deutonymphs of the mites clung onto the beetle larvae and became enclosed in the pupal cells so that newly emerged adult beetles already had a mite population to carry to a new corpse.

Disease organisms might affect competition in a number of ways. Vizoso (1969) suggested that the stress caused by competition between red squirrel (*Sciurus vulgaris*) and grey squirrels (*S. carolinensis*) increased the virulence of virus disease. This might have been a factor in the replacement of red by grey in Britain, but the red was already subject to large fluctuations before the grey was introduced (Chapter 7). It is also possible that new diseases may be introduced with closely related species and the pre-existing potential competitor is reduced by them rather than by competition.

References

Al-Mufti, M. M., Sydes, C. L., Furness, S. B., Grime, J. P. and Band, S. R. (1977). A quantitative analysis of shoot phenology and dominance in herbaceous vegetation. *J. Ecol.* **65**, 759–91.

Buss, L. W. (1979). Bryozoan overgrowth interactions – the interdependence of competition for space and food. *Nature* **281**, 475–7.

Catchpole, C. K. (1973). Conditions of coexistence in sympatric breeding populations of *Acrocephalus* warblers. *J. Anim. Ecol.* **42**, 623–35.

Catchpole, C. K. (1978). Interspecific territorialism and competition in *Acrocephalus* warblers as revealed by playback experiments in areas of allopatry and sympatry. *Anim. Behaviour* **26**, 1072–80.

Clark, P. and Evans, F. (1954). Distance to nearest neighbour as a measure of spatial relationships. *Ecology* **35**, 445–53.

Cody, M. L. (1969). Convergent characteristics in sympatric species: a possible relation to interspecific competition and aggression. *Condor* **71**, 222–39.

Connell, J. H. (1961). The influence of interspecific competition and other factors on the distribution of the barnacle *Chthalamus stellatus. Ecology* **42**, 710–23.

Dagg, A. I. and Foster, J. B. (1976). *The Giraffe its Biology Behaviour and Ecology.* New York: Van Nostrand Reinhold.

Donisthorpe, H. St J. K. (1927). *British Ants their Life-History and Classification.* London: Routledge.

Edington, J. M. and Edington, M. A. (1972). Spatial patterns and habitat partition in the breeding birds of an upland wood. *J. Anim. Ecol.* **41**, 331–57.

Fisher, R. C. (1961). A study in insect multiparasitism. II. The mechanism and control of competition for possession of the host. *J. Exp. Biol.* **38**, 605–28.

Fricke, K. (1904). "Licht und Schattenholzarten": ein wissenschaftlich nicht begründetes Dogma. *Centralbl. f. d. gesamte Fortwesen* **30**, 315–25.

Gordon, D. P. (1972). Biological relationships of an intertidal Bryozoan population. *J. Nat. Hist.* **6**, 503–14.

Grant, P. R. (1972). Interspecific competition among rodents. *Ann. Rev. Ecol. Syst.* **3,** 79–106.

Grant, V. and Grant, K. A. (1965). *Flower pollination in the Phlox family.* New York: Columbia University Press.

Grime, J. P. (1979). *Plant Strategies and Vegetation Processes.* Chichester, New York: Wiley.

Harper, J. L. (1977). *Population Biology of Plants.* London, New York: Academic Press.

Heinrich, B. and Raven, P. H. (1972). Energetics and pollination ecology. *Science* **176,** 597–602.

Hubbell, S. P. and Johnson, L. K. (1977). Competition and nest spacing in a tropical stingless bee community. *Ecology* **58,** 949–63, (The Ecological Society of America).

Iverson, S. L. and Turner, B. N. (1972). Winter coexistence of *Chlethrionomys gapperi* and *Microtus pennsylvanicus* in a grassland habitat. *Am. Mid. Nat.* **88,** 440–5.

Jackson, L. B. C. and Buss, L. (1975). Allelopathy and spatial competition among coral reef invertebrates. *Proc. Nat. Acad. Sci. USA* **72,** 5160–3.

Karli, P. (1956). The Norway rat's killing response to the white mouse: an experimental analysis. *Behaviour* **10,** 81–103.

Krämer, A. (1973). Interspecific behaviour and dispersion of two sympatric deer species. *J. Wildl. Manage.* **37,** 288–300.

Lanyon, W. E. (1956). Territory in the meadowlarks, genus *Sturnella. Ibis* **98,** 485–9.

Levin, D. A. and Anderson, W. W. (1970). Competition for pollinators between simultaneously flowering species. *Amer. Naturalist* **104,** 455–67.

Lofqvist, J. (1977). Toxic properties of the chemical defense systems in the competitive ants *Formica rufa* and *Formica sanguinea. Oikos* **28,** 137–51.

Majer, J. D. (1972). The ant mosaic in Ghana cocoa farms. *Bull. Ent. Res.* **62,** 151–60.

Majer, J. D. (1976). The maintenance of the ant mosaic in Ghana cocoa farms. *J. Appl. Ecol.* **13,** 123–44.

Menge, B. A. (1976). Organisation of the New England rocky intertidal community: role of predation, competition, and environmental heterogeneity. *Ecol. Monogr.* **46,** 355–93.

Milner, J. S. (1976). Effects of food deprivation and competition on mouse killing in the rat. *Bull. Psychon. Soc.* **7,** 442–4.

Moore, N. W. (1964). Intra- and inter-specific competition among dragonflies. *J. Anim. Ecol.* **22,** 253–73.

Morse, D. H. (1974). Niche breadth as a function of social dominance. *Amer. Naturalist* **108,** 818–30.

Moxon, J. E. (1980). *Comparative Analysis of the Interspecific Aggressive Behaviour of some British Ants with Particular Reference to Myrmica spp. and Lasius flavus* (F.). Ph.D. Thesis, London University.

Murray, B. G. (1971). The ecological consequences of interspecific territorial behaviour in birds. *Ecology* **52,** 414–23.

Murray, B. G. (1976). A critique of interspecific territoriality and character convergence. *Condor* **78,** 518–25.

Nicholson, A. J. (1955). An outline of the dynamics of animal populations. *Austral. J. Zool.* **2,** 9–65.

Pickles, W. (1944). Territories and interrelations of two ants of the genus *Messor* in Algeria. *J. Anim. Ecol.* **13**, 128–9.

Pontin, A. J. (1960). Field experiments on colony foundation by *Lasius niger* (L.) and *L. flavus* (F.) (Hym., Formicidae). *Insectes Sociaux* **7**, 227–30.

Pontin, A. J. (1961). Population stabilization and competition between the ants *Lasius flavus* (F.) and *L. niger* (L.). *J. Anim. Ecol.* **30**, 47–54.

Primack, R. B. and Howe, H. F. (1975). Interference competition between a hummingbird (*Amazilia tzacatl*) and skipper butterflies (Hesperiidae). *Biotropica* **7**, 55–8.

Purcell, J. E. (1977). Aggressive function and induced development of catch tentacles in the sea anemone *Metridium senile* (Colenterata, Actinaria). *Biol. Bull.* **153**, 355–68.

Salt, G. (1961). Competition among insect parasitoids. *Symp. Soc. Exp. Biol.* **15**, 96–119.

Selander, R. K. and Giller, D. R. (1959). Interspecific relations of woodpeckers in Texas. *Wilson Bull.* **71**, 107–24.

Shields, W. M. and Bildstein, K. L. (1979). Birds versus bats: behavioural interactions at a localised food source. *Ecology* **60**, 468–74.

Simmons, K. E. L. (1951). Interspecific territorialism. *Ibis* **93**, 407–13.

Slagsvold, T. (1978). Competition between the great tit *Parus major* and the pied flycatcher *Ficedula hypoleuca*: an experiment. *Ornis Scand.* **9**, 56–50.

Springett, B. P. (1968). Aspects of the relationship between burying beetles, *Necrophorus* spp. and the mite, *Poecilochirus necrophori* Vitz. *J. Anim. Ecol.* **37**, 417–24.

Stimson, J. (1970). Territorial behaviour of the owl limpet, *Lottia gigantea*. *Ecology* **51**, 113–8.

Turner, B. N., Perrin, M. R. and Iverson, S. L. (1975). Winter coexistence of voles in spruce forest: relevance of seasonal changes in aggression. *Can. J. Zool.* **53**, 1004–11.

Vanderplank, F. L. (1960). The bionomics and ecology of the red tree ant *Oecophylla* sp., and its relationship to the coconut bug *Pseudotheraptus wayi* Brown (Coreidae). *J. Anim. Ecol.* **29**, 15–33.

Vizoso, A. D. (1969). A red squirrel disease. *Symp. Zool. Soc. Lond.* **24**, 29–38.

Vroey, C. de (1979). Aggression and Gause's law in ants. *Physiol. Entomol.* **4**, 217–22.

Waser, N. M. and Real, L. A. (1979). Effective mutualism between sequentially flowering plant species. *Nature* **281**, 670–2.

Way, M. J. (1953). The relationship between certain ant species with particular reference to biological control of the coreid, *Theraptus* sp. *Bull. Ent. Res.* **44**, 669–91.

Went, F. W. (1942). The dependence of certain annual plants on shrubs in Southern California deserts. *Bull. Torrey Bot. Club* **69**, 100–14.

Williams, R. B. (1975). Catch tentacles in sea anemones: occurrence in *Haliplanella luciae* (Verrill) and a review of current knowledge. *J. Nat. Hist.* **9**, 241–8.

Wilson, E. O. (1971). *The Insect Societies*. Cambridge, Mass.: Belknap.

Wright, H. O. (1973). Effect of commensal hydroids on hermit crab competition in the littoral zone of Texas. *Nature* **241**, 139–40.

4 Coexistence and competition

4.1 The conditions for coexistence

There are several ways in which species can share resources for as long as the habitat remains unchanged and they need to be distinguished at the beginning of this section. Not all of them involve competition in the sense used here and therefore it must be important to provide evidence of population reduction to establish that competition is taking place.

(a) Interspecific competition may be stabilized intrinsically because two species are bound to differ in exploitation. This has already been discussed briefly in the Introduction, but experience has shown this to be an unexpectedly difficult point to explain in the face of false dogma. Another very simple argument will therefore be presented here. For the sake of simplicity consider only two species exploiting a range of food in such a way that part of the range is shared by chance, part is more likely to be eaten by A and the rest more likely to be eaten by B. The food is renewed at a fixed rate and it is all consumed. More individuals of A immigrate into the mixed populations. These individuals will take the same range of food as the A already present, but only some of the range of food taken by the B individuals. The subsequent survival of A will therefore be less than that of B thus automatically counteracting the immigration. Even allowing for intraspecific variation, individuals of different species will differ more from each other than individuals of the same species so that coexistence will be potentially stable in the absence of interference. The actual details of resources, the methods of exploitation or the degree of overlap do not matter provided that this simple maxim holds good.

Two other points are worth underlining in this context. More closely related species are not necessarily more likely to displace each other and displacement could be rendered less likely if closely related species are similarly affected by factors which cause change in numbers and might otherwise be expected to cause departure from equilibrium proportions.

(b) Interference is the main cause of displacement of one species by another and this is the subject of the next chapter, but interference may be reduced in two ways. Many differences between species may improve stability because they reduce the frequency of meetings. Stratification in space or by temporal differences in activity should reduce interference and promote coexistence even if the species still share the same population of food.

Secondly, where population densities are kept very low by other factors, such as predators and parasites of all sorts, interference may again be too rare to be significant in destabilizing coexistence. Superabundance of resources is usually considered to be important for coexistence under these circumstances, but if exploitation is stabilizing coexistence then superabundance of resources is irrelevant. Similarly, independent control of numbers of species by separate factors has frequently been put forward to explain the prediction of coexistence by Lotka–Volterra models, but this is unnecessary.

(c) Predators may kill a disproportionately large number of whichever species is most abundant and this behaviour may contribute to stable coexistence. Section 4.13 discusses predator and herbivore effects further with examples.

(d) Diversity may be maintained by a 'dominance ring' of species as suggested by Jackson and Buss (1975). Their hypothesis requires species A to displace B which in turn displaces A, giving continual changeover of species in the community. Stability from some other cause may well be necessary.

(e) Fugitive species with superior powers of dispersal may possibly coexist with species having superior growth (Skellam, 1951). Death of individuals of the dominant species leaves gaps which are more quickly filled by the fugitive species and this must colonize further gaps before it is displaced by the dominant.

(f) Mosaics of species may result from the chance pattern of colonization and if the mature individuals prevent further colonization by any of the codominants then the mosaic may persist. Ant colonies or trees may provide examples as discussed in 3.2.

(g) Three more rather spurious types of 'coexistence' should be borne in mind. Mosaics may also be produced by an underlying habitat mosaic which has been overlooked by the investigator and:

(h) The displacement of one species by another may be very slow and if observed in transition may be overlooked. Evidence of return to equilibrium after experimental disturbance may be needed to discount this possibility.

(i) A 'coexisting' mixture of mobile species may be observed in a zone between their preferred habitats into which they are continually diffusing.

In view of all these alternative possibilities for explaining coexistence three main criteria are needed to make an example of research relevant to this chapter on stable competition. They are: (1) evidence that one species is reducing the numbers of the other and, ideally, this should be reciprocal. The evidence needs to be experimental and under as near to normal field conditions as possible with adequate controls. (2) Evidence that the quantity of shared resources is important in determining the population sizes of the competing species. (3) Evidence of return to equilibrium after disturbance of the relative proportions of the competing species.

The difficulty of working with field populations means that no one example includes all these features, but the single most important criterion is the performance of experiments. Ecology still lags behind most branches of science in the use of standard scientific method, probably because ecologists have been too easily diverted from field work into flights of theorizing or into studies of individual behaviour or physiology.

There are nevertheless a number of examples available and these are reviewed next, starting with an illuminating laboratory study.

4.2 *Tribolium* versus *Oryzaephilus* in laboratory cultures

Crombie (1945, 1946) performed extensive experiments with grain beetles in cultures and these have been too frequently overlooked. His populations were of course very artificial and lacked many of the complications of resource variety or interaction with other species which are characteristic of more natural communities. This simplicity makes some of the experiments well worth reviewing as an introduction to the theory used and should help to forestall illogical argument about ideas which are really very simple.

A number of species were tried in various combinations, but the most illuminating series of experiments concerns *Tribolium confusum*, a 3-mm-long tenebrionid very commonly studied in cultures, and a slightly smaller beetle *Oryzaephilus surinamensis*. They both exploit the same food which can be wheat either as whole grain or ground into flour and in the experiments described here this is renewed each week. They could also be regarded as competing for the confined space of the culture jar. Interference is also very important because they are cannibalistic. *Tribolium* adults and larvae eat eggs and pupae of their own species or those of *Oryzaephilus*. Adult *Oryzaephilus* destroy *Tribolium* eggs, but at a lower rate than its own are destroyed by *Tribolium*. They also eat their own eggs and pupae. *Tribolium* also has another advantage as it has higher fecundity under these conditions and can potentially easily outstrip *Oryzaephilus* if cultured with it. Contacts between individuals increase in frequency with density and population growth ceases when eggs and pupae are found as fast as they are produced.

Consider culture in cracked wheat grain only, to start with, and in this case it was found that either species alone increased in a standard culture vessel to about 420–450 adults in 150 days and this density was maintained to the end of the experiment at 245 days. It has become conventional to call this maximum level the carrying capacity (K) of the environment.

When cultured together they both persist for at least as long, but *Tribolium* reaches an adult number of about 360 and *Oryzaephilus* tends to oscillate between 100 and 200. The initial proportions of the species used to start the compeition are not important to the outcome and Crombie showed a similar equilibrium mixture starting from densities of 4 of each, 100:4 or 4:100, and

4 *Tribolium* to 400 *Oryzaephilus* (Fig. 4.1). Both species are depressed in numbers by the presence of the other and *Oryzaephilus* is worse affected. There is no question that they are competing and stability is shown by the convergence onto one equilibrium mixture from a range of widely different mixtures. The total number of beetles in mixed culture at equilibrium is equal to or greater than the carrying capacity number of either alone so the combination may be more efficient at converting grain to beetles.

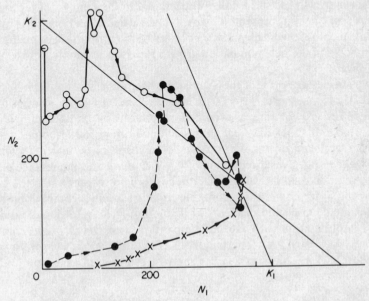

Figure 4.1 The progress of three competition experiments started with different numbers of *Tribolium* (N_1) and *Oryzaephilus* (N_2) in cracked wheat cultures. Lines of zero growth of each species are drawn using Lotka/Volterra theory (*cf.* Fig. 4.2). (*From* Crombie, 1946.)

If the culture medium is changed to fine flour the adult maximum for *Tribolium* alone is about 135, and 120 for *Oryzaephilus* alone. Attempts to culture the two together fail in this medium and *Oryzaephilus* declines rapidly to extinction from 35 days to 250 days. Extinction would be much quicker if the adults did not take a long time to die of old age as the immature *Oryzaephilus* are quickly killed. A large superiority in number of *Oryzaephilus* slows down the increase of *Tribolium*, but the eventual result is the same. Interference has clearly become one-sided and immatures are more exposed to attack in this food so this aspect was investigated further.

Experimental cultures were then set up with 30 pieces of narrow glass tubing each 2 cm long added to each culture. If this tubing was 1-mm bore it allowed *Oryzaephilus* to pupate inside, but adult or large *Tribolium* larvae could not enter. A stable mixture of the species was produced in competition experiments with this protective shelter available for *Oryzaephilus*. The equilibrium numbers are different from the wheat culture at 175 adult *Tribolium* and 80 adult *Oryzaephilus*. If the tubing has 2-mm bore then all stages of both species can enter, *Tribolium* wins and does better alone if its own pupae are protected by this size of shelter.

Lotka/Volterra models are easily applied to these results. Consider first a culture with only one beetle species which increases to a maximum number K. Population growth rate is initially proportional to the number of reproducing adults N and can be represented by rN in uncrowded conditions. The rate will be reduced with crowding by a factor related to the proportion of medium still unoccupied and the new population growth rate is most simply represented as $rN(K - N)/K$.

If two species are present they will affect each others' growth rates, but an individual of one species, say *Tribolium*, will not have the same effect on *Oryzaephilus* growth rate as would be produced by another individual of *Oryzaephilus*. In cracked wheat culture at equilibrium 360 *Tribolium* result in 300 less *Oryzaephilus* and 150 *Oryzaephilus* result in 60 fewer *Tribolium* – one *Oryzaephilus* gives a reduction of 0.4 *Tribolium* and the reverse effect is 0.8 : 1. An additional crowding effect needs to be incorporated in the growth rate formula to convert it from purely intraspecific competition to an interspecific competition model. For *Tribolium* $(r_1 K_1 N_1)$ it looks intuitively that its population growth rate in competition with *Oryzaephilus* should be $rN_1(K_1 - N_1 - 0.4N_2)/K_1$. The factor for converting species two into species one in terms of its competitive effect may be symbolized as α_{12} and the basic equation becomes:

$$\frac{dN_1}{dt} = r_1 N_1 \left(\frac{K_1 - N_1 - \alpha_{12}N_2}{K_1} \right)$$ and similarly for the second species:

$$\frac{dN_2}{dt} = r_2 N_2 \left(\frac{K_2 - N_2 - \alpha_{12}N_1}{K_2} \right)$$

At equilibrium, if it occurs,

$$\frac{dN_1}{dt} = 0 = \frac{dN_2}{dt}$$

The culture is then saturated so $K_1 - N_1 - \alpha_{12}N_2 = 0$ and $K_2 - N_2 - \alpha_{12}N_1 = 0$. Therefore $\alpha_{12} = (K_1 - N_1)/N_2$ and $\alpha_{21} = (K_2 - N_2)/N_1$ which we have already deduced.

Not all values of K and α will give prediction of stability and a graphical method of representing the models (see Crombie, 1945; Williamson, 1957; Krebs, 1972) provides a convincing way of visualizing this. If two axes are used to represent the population sizes of the two species as in Fig. 4.2 then one can immediately put on two points of zero growth for each species. When $K_1 - N_1 - \alpha_{12}N_2 = 0$, then $N_1 = K_1$ when $N_2 = 0$, and $N_2 = K_1/\alpha_{12}$ when $N_1 = 0$, for species one.

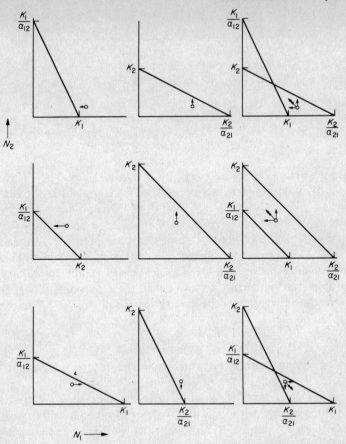

Figure 4.2 The three possible outcomes of competition between two species. The left-hand column shows zero growth lines for species N_1 with its direction of change in number from the starting mixture shown. The centre column is the same treatment, but for species N_2 only. On the right the two species are superimposed to show the resultant direction of change of the starting mixtures. The top row models stable competition, the middle row shows unstable competition with species 2 always displacing species 1 and the bottom row also shows unstable competition, but either species could win.

A line connecting these two points is a locus of all points with zero growth of species one (or similarly for species two). Starting densities of N_1 will approach the zero growth line as indicated by the arrows, whatever the value of N_2, and, since the axes are the same, the diagrams for the two species may be superimposed. A starting mixture of the two species together will now be seen to grow together in a direction determined by the vectors of N_1 and N_2.

For *Tribolium* and *Oryzaephilus* in cracked wheat $K_1 = K_2$ and both α

Table 4.1 The numbers of adult *Tribolium* and *Oryzaephilus* competing in renewed media (From Crombie, 1946.)

	Equilibrium nos.								
	Alone		Together						
Medium	K_1	K_2	N_1	N_2	α_{12}	α_{21}	K_1/α_{12}	K_2/α_{12}	Results
Cracked wheat	425	445	360	150	0.4	0.8	531	1113	Stable competition
Fine flour with	175	400	175	80	small	1.8	large	222	Stable, *Or.* has little
1-mm tubes									effect on *Trib.*

values are less than one, therefore $K_1/\alpha_{12} > K_2$ and $K_2/\alpha_{21} > K_1$. This arrangement leads to an equilibrium mixture at the crossing point of the zero growth lines.

If either $K_1/\alpha_{12} < K_2$ or $K_2/\alpha_{21} < K_1$ or both, then instability will result as shown in cases 2 and 3 of Fig. 4.2. One species is completely superior to the other in case 2 and could easily be achieved by interference as with *Tribolium* and *Oryzaephilus* in fine flour. Case 3 could result in either winning depending on starting density or r values in the environment concerned.

For more than two species the general equation is:

$$\frac{dN_i}{dt} = N_i r_i \left(\frac{K_i - N_i - \Sigma \alpha_{ij} N_j}{K_i} \right)$$

A matrix of alphas is produced which could be used in analysing a guild of species.

In the field considerable practical difficulties in applying these models arise. K values are measurable if both species can be allowed experimentally to increase to a maximum by removal of the other in separate areas, but naturally occurring populations of one in the absence of the other may well have other differences which affect K. Alpha values also need to be measured directly by experiment. Simple measurement of overlap of resources used might measure alpha, but it might not if the populations are limited by factors other than the resources measured. It is necessary to test whether the species are interacting first and in any case alpha is a compound of exploitation and interference needing caution in measuring resources used rather than wasted.

4.3   Ants

Social life with well-defined nest sites results in populations which can be very conveniently altered for field experiments, but two problems arise when using ants to investigate phenomena also shown by more typical animals. Each nest is best thought of as one individual, so therefore the young reproductives produced represent its fecundity, and the nest size is so variable that a population of nests needs more than a simple count of nests to describe its density. Secondly, it may appear that the nests are static and the population rather plant-like, but most ants can move their nests for distances similar to the normal foraging range. This means that nests can rearrange themselves as they grow to fill vacant spaces in the population or exploit special local conditions.

In Britain two species of *Lasius* are abundant in grassland and normally occur together. *L. flavus* makes grass-covered mounds with a regular spacing of 2–3 m in old grazed grassland and is normally subterranean with corresponding pale colour and reduced eyes. *L. niger* forages above and below ground resulting in a much larger spacing between nests of around 8 m (Fig. 4.3). These species are therefore very different for congeners and have been placed in separate genera by some authors (e.g. Donisthorpe, 1927). Their differences and similarities are listed in Table 4.2.

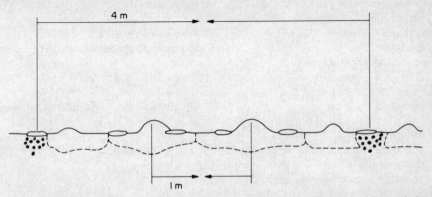

Figure 4.3 Diagrammatic vertical section of a stony grassland habitat with coexisting *Lasius flavus* in mound nests and *L. niger* nesting under superficial stones. The population of *L. niger* has large (4 m radius) intraspecific territories above ground and is superimposed on the population of *L. flavus* which has intraspecific territories (1 m radius) entirely below ground. (*From* Pontin, 1961.)

The facts that they normally occur together and have obvious differences could lead a 'competitive-exclusionist' to assume they were not competing because they had partitioned the habitat and achieved ecological isolation. There are many similar assumptions in the literature, but in this case at least they are far from the truth. Populations of nests were mapped and their alate queen production counted to assess population size (Pontin, 1961a, 1963, 1969). Direct tests of both intra- and interspecific interaction were made by experimentally eliminating *L. niger* and transplanting *L. flavus* nests in the field to measure the resulting changes in queen production compared with controls.

One *L. niger* nest which had produced 17.7 g of queens (687) in 1956 was removed using poisoned food in March 1957. The *L. flavus* nests here

Table 4.2   The differences and similarities between *Lasius flavus* and *L. niger*

Differences

L. niger	*L. flavus*
Dark black/brown	Yellow/pale brown
Normal compound eyes	Reduced compound eyes
Worker size uniform in one nest	Worker size often variable in same nest
Forages above ground and below	Forages below ground only
Intraspecies territory radius 4 m	Intraspecies territory radius 1 m
More active and faster running	Slower at the same temperature
Nests are loose earth mounds or under stones	Nestmounds permanent and consolidated by plants
Successful in many open habitats	Grazed grassland main habitat
Aphid eggs not cared for	Aphid eggs treated like ant eggs
Aphid enemies attack aphids in nests	Aphids in ant tunnels not attacked
Aphids commonly produce alatae in nests	Aphid alatae rare
Alate queen live wt 0.026 g	Alate queen wt 0.0175 g
Monogynous	Frequently polygynous at any age of colony
Mandibular gland secretions: 1-octanol, 1-nonanol	Citronellal, 2,3-dihydrofarnesal
Dufour's gland secretions: many aliphatic acetates	4-hydroxyoctadec and related substances

Similarities
  Tend the same range of subterranean myrmecophilous aphids
  Same seasonal cycle and life-history timing
  Select the same habitats after mating flights
  Prey on myrmecophilous aphids

produced 7.6 g (434) more queens in summer 1957 compared with 1956 and the control queen production was similar in the two years. A few adult *L. flavus* workers are occasionally recorded in the food of *L. niger* (Pontin, 1961*b*), but predation cannot be an important cause of this interaction. Sufficient *L. flavus* workers must have been present in March to rear the increased queen number because replacement workers would not become adult before June when the queen larvae were fully grown anyway.

Almost half the food of *L. niger* would need to come from below ground to produce this effect and it may be a higher proportion if more than one season is needed for *L. flavus* to recover fully. On the other hand, *L. flavus* queen production averages $0.25\,\text{g\,m}^{-2}$ in this area with *L. niger*, $0.31\,\text{g\,m}^{-2}$ where *L. niger* is absent. *L. niger* averages $0.13\,\text{g\,m}^{-2}$ and was not found without *L. flavus* in the research area. This *L. niger* production therefore coincides with $0.06\,\text{g\,m}^{-2}$ less *L. flavus* which would be inadequate evidence alone, but it backs up the previous measurement of competitive effect extremely well. α is estimated to be 0.43.

As well as the primary effect of this experiment repercussions on other members of the populations were recorded. Neighbouring nests of *L. niger* showed increased queen production and the *L. flavus* nests close to them showed a corresponding reduction.

The transplantation experiment (Pontin, 1969) was performed to test the reciprocal effect of *L. flavus* on *L. niger*. *L. flavus* nest-mounds were dug out as intact as possible and moved away from some *L. niger* nests to sites near to other *L. niger* nests. It was shown that *L. flavus* could severely reduce *L. niger* production if large enough numbers were added, but the normal effect was probably small and the *L. niger* recovered in two years. Also it was clear that *L. flavus* had much more effect on other *L. flavus* than on *L. niger* as predicted by exploitation competition theory (Table 4.3). Removal of *L. flavus* nests producing 48.5 g of queens was followed by an increase of 14.6 g in *L. niger* queen production which suggests that the α for *L. flavus* effect on *L. niger* could be as high as 0.3.

Table 4.3 Maximum changes in queen production following transplantation of 6 nests of *L. flavus* (From Pontin, 1969.)

		Effect on queen production by neighbouring nests				
		L. niger			L. flavus	α
	No. of nests	*Total queen production*		*No. of nests*	*Total queen production*	
Treatment		*No.*	*Biomass, g*		*No.*	*Biomass, g*	
Removal of *L. flavus* producing 47 g of queens	2	+560	+14.5	5	+1378	+24	0.3
Addition of *L. flavus* producing 47 g of queens	2	−270	−7	7	−1093	−19	

Evidence of the importance of shared resources was also obtained in the same studies. *L. flavus* had to be tested indirectly for territorial limitation of queen production because of its subterranean behaviour. Only two measurements were available: distance between nest-mounds and number of queens produced, but if the territories were near enough circular and concentric around the nests then the equation $\sqrt{N_1} + \sqrt{N_2} = D$ (where N_1 and N_2 are numbers of queens produced by neighbouring nests distance D apart) would be true if the area were saturated with territories. The youngest site (17 yr from disturbance) where many small nests were found filling the gaps showed a good fit, but the older site (>30 yr) was less good.

L. niger nests were regularly spaced out by their large above-ground territories, but too few were available for further analysis. Far more superficial stones were available for nest sites than were used by this species in the young site. Both species, therefore, show territorial behaviour and the evidence for the size of feeding territory determining queen production is very strong for *L. flavus*.

The food actually taken in their territories is difficult to observe, but samples can be collected from the masses of larvae in the nest centres (Pontin, 1961*b*). *L. niger* is very much a 'Jack-of-all-trades' whereas *L. flavus* is a very highly adapted aphid farmer (Pontin, 1978) eating a much higher proportion of its myrmecophilous aphids in its prey, suppressing alate production by its aphids, storing aphid eggs (Pontin, 1961*c*) and protecting them from predators and parasites (Pontin, 1960).

L. niger tends the same species of aphids and also any available on the above-ground parts of plants, but in open grassland the only myrmecophilous aphids available are subterranean species. *L. niger* has not been recorded to store aphid eggs, alates are produced freely by aphids in its tunnels and many myrmecophilous aphid enemies attack its enclosed aphids (Pontin, 1960).

Interference, both intra- and interspecific, is usually confined to fighting in spring. The battles between neighbouring *L. niger* nests may involve thousands of workers for several days in succession and hundreds of corpses appear as food in the nests at this time only. Battles between *L. flavus* and *L. niger* were not observed at the young and very stony site, but *L. niger* is able to displace *L. flavus* from hot sites in old pasture where superficial stones are rare. *L. flavus* merely starts a new nest-mound next to the site lost and was not observed to die out when in association with *L. niger*. Perhaps the chemical defences of *L. flavus* are effective (*see* Bergström and Löfqvist, 1970) as relatively few are found as food in *L. niger* nests. It is very rare to find corpses of *L. flavus* in nests of its own species and intraspecific battles in this species must be correspondingly rare. Its nests are, nevertheless, distinct. No transfer of radioactive tracer from one nest to another was found by Odum and Pontin (1961).

The founding queens of both species were found to be killed by nests of either (Pontin, 1960) species, but queens near nests of the opposite species survived longer which may help to ensure a mixture in newly bared sites which both species chose for colonization.

The two species have been observed to coexist for more than 30 yr, but these are the same individual nests so the best evidence for stability is provided by the experiments. Disturbance of the populations is followed by return to equilibrium and not unstable change towards one species or the other. This is particularly obvious in the transplant experiment because of the relatively high speed of the recovery – the effects of adding extra *L. flavus* nests were produced one year later, but recovery to the original density was clearly taking place two years later (Pontin, 1969).

4.4   Great tit and blue tit

The genus *Parus* includes nine European species of small tree-feeding passerine birds. Two of these, *P. major* and *P. caeruleus*, have been most extensively studied by many authors and are normally found together in the same broadleaved woodland.

Hartley (1953) recorded feeding positions of five species of *Parus* and showed both a high degree of overlap in blue and great tit feeding distribution and also some characteristic differences (Fig. 2.2). Great tit fed at lower levels except during the spring outbreak of defoliating caterpillars when both concentrated on these and the great tit fed more on the ground, especially in the winter where beechmast was available. Winters with long periods of snow cover may be more severe on the great tit as result of this (*see* Perrins, 1979).

Betts (1955) analysed the gizzard contents of four *Parus* species and confirmed overlapping exploitation of food resources with characteristic differences between blue and great tits. This author concluded that there was 'little similarity in the diets of the great and blue tits, the great tit taking mainly adult insects, especially weevils, while the blue tit fed mainly on scale insects and small larval and pupal forms'. But, there was very extensive overlap in the breeding season (Fig. 4.4) when the larvae and pupae of oak tortrix and winter moth form a high proportion of the food of both species.

The results of the Ghent study presented by Dhondt (1977) are therefore of

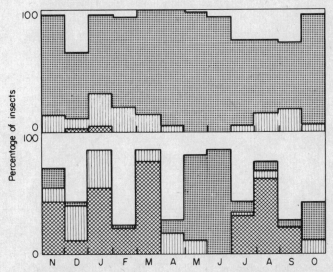

Figure 4.4   Percentage of insects taken by blue (upper) and great tits for each month from various habitats: the ground layer is cross-hatched, trunks and branches vertical hatched, twigs and leaves stippled. The remainder is indeterminate. (*From* Betts, 1955.)

particular interest. This study was made in deciduous woodland with excess nestboxes which were used by nearly all the pairs of these species present. Stable coexistence is clearly shown, in the face of disturbing winter mortality, by the relative numbers of the two species changing towards a central equilibrium mixture in almost all the 16 years (Fig. 4.5). The main exception is 1962–3 which was an exceptionally severe winter. This stability is expected from the stronger correlations between breeding success and each species' own density compared with correlations between breeding success and the density of the other species (Table 4.4). A direct causal effect of each species on the other is probable, but further work is needed to confirm this.

Gibb (1954) noted frequent supplanting attacks by feeding tits and in the case of blue and great tit over 90% were attacks by a member of the same species. In winter these attacks averaged 5 times per hour in the great and 15

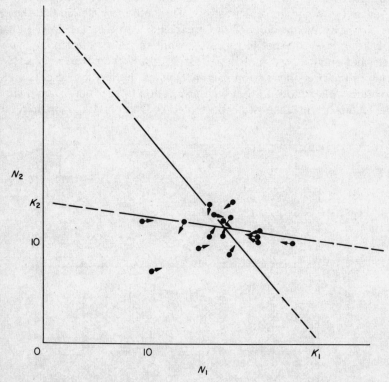

Figure 4.5 The densities of great (N_1) and blue (N_2) tits per 10 ha in coexisting populations for each of the years 1960–76. The direction of change to reach the succeeding year's figures is indicated in each case. The zero-growth lines are drawn so that the changes fit Lotka/Volterra theory (*c.f.* Fig. 4.2). Only one point cannot be fitted and this is for the exceptionally severe winter of 1962–3. (*From* Dhondt, 1977, reproduced here by courtesy of *Nature*.)

Table 4.4   Correlations between breeding success and numbers of the same species or numbers of different species

| | Density of great tit | | Density of blue tit | |
	Pairs/10 h	P	Pairs/10 h	P
Great tit productivity, fledglings/pair/year	−0.69	<0.005	−0.64	<0.005
Blue tit productivity fledglings/pair/year	−0.39	>0.05	−0.46	<0.05

From the 16-year Ghent study of blue tit and great tit (Dhondt, 1977, reproduced here by courtesy of *Nature*.)

times per hour in the blue so they may well be important at a time when food is scarce. Dominance of the larger great tit was not clear cut although it may sometimes displace the blue tit from a nest site (*see* Perrins, 1979, for a review). Interference could in this case contribute stability since intraspecific attacks exceed interspecific attacks.

Dhondt attempted to estimate K and alpha values by fitting lines of zero growth to each species as usual when applying Lotka/Volterra modelling. It is certainly possible to draw lines for each species (Fig. 4.5) with points showing subsequent increase below, no change on them, and the others above. This gives an equilibrium mixture of 1.7 prs/h great tit and just over 1.2 prs/h of blue tit. K for great tit is around 3 prs/h and α for converting blue to great is estimated at 1.2. For blue tit K is 1.3 prs/h and the reciprocal α is 0.13. This assumes that such a simple model fits and it would be better to estimate alphas and Ks by experimental removal of each species if practicable.

4.5   Jackdaw and magpie

Högstedt (1980) reported an interesting experiment to test the effect of Jackdaw (*Corvus monedula*) upon the magpie (*Pica pica*) by introducing nestboxes into magpie habitats which lacked breeding sites for the jackdaw. The experiment was performed in two years and the experimental magpies in one year were used as controls for the other year. Jackdaws readily occupied the nestboxes and the magpies subjected to added jackdaw populations showed reduced breeding success compared with controls (Table 4.5).

The ratio of reduction of biomass of magpie fledglings to increase in biomass of jackdaws was estimated to be 0.43 and is an alpha value for converting jackdaw to magpie in a competition equation if equilibrium has been reached.

The causes of this interaction may be complex. Food overlap, estimated from samples brought to nestlings by parents, was 0.54 and some nestlings

Table 4.5 Breeding success of magpies with experimentally increased jackdaw competition compared with controls

	Experimental group	Controls
Clutch size . . .	6.5 ± 0.3	6.48 ± 0.18
No. of started breedings . . .	18	38
No. of hatched clutches . . .	10	23
No. of broods with one or more fledged . . .	3	16
No. of fledged young per breeding attempt . . .	0.33 ± 0.2	1.68 ± 0.34

The difference in fledgling production has a $P < 0.001$. (From Högstedt 1980, reproduced here by courtesy of *Nature*.)

apparently starved so exploitation competition could be the main explanation. Interference was also observed once the young were hatched and 101 attacks on foraging jackdaws by magpies were recorded in 35 h observation. Some nestlings were preyed on by crows (*Cornix cornix*) and hungry nestlings beg noisily so predation and competition effects are not independent if crows use nestling sounds to find their prey.

The reciprocal effect of magpies on jackdaws is not so easily assessed, but if exploitation is important it could be of similar magnitude and it would be valuable to have some measurement of it.

4.6   Starfish

Transfer experiments in the field have been performed on *Pisaster ochraceus* which coexists with *Leptasterias hexactis* (Menge, 1972; Menge and Menge, 1974).

Pisaster is much larger than *Leptasterias* and reaches 550 g wet weight although it is reproductively mature 70–90 g. *Leptasterias* only attains 11.5 g and can reproduce at 2 g. Biomass per m² was shown to be inversely correlated for the two species at a number of sites (Fig. 4.6) and to test whether this was caused by interaction *Pisaster* was collected from one reef and transferred to another. A year after the transfer was started *Leptasterias* had increased from 2.0 to $4.2 \mathrm{g\,m^{-2}}$ where $110 \mathrm{g\,m^{-2}}$ *Pisaster* had been effectively removed and decreased from 4.3 to $3.5 \mathrm{g\,m^{-2}}$ where *Pisaster* had been added to produce an increase of from 100 to $230 \mathrm{g\,m^{-2}}$ in its own density. More results from controls are needed.

No temporal or spatial differences between the species were recorded when feeding and there was a very large overlap in the range of prey species taken. *Leptasterias* tended to prey more on mobile molluscs while *Pisaster* showed some preference for sessile molluscs and barnacles or larger specimens of the same taxa taken by *Leptasterias*.

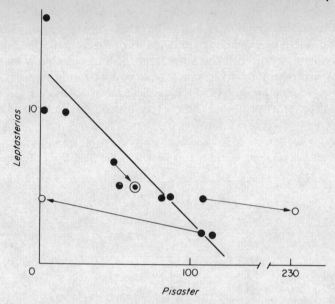

Figure 4.6 Density (g/m²) of the starfish *Pisaster* (abscissa) and *Leptasterias* for a number of sites. Changes following experimental transfer of *Pisaster* are shown by the two open circles and the double circle is a control. (*From* Menge, 1972.)

Interference was important as *Pisaster* attacked *Leptasterias* using pedicillariae and was actively avoided. The smaller *Leptasterias* could enter smaller crevices and this may be the important factor in allowing coexistence in the face of *Pisaster* dominance. Interestingly, small individuals of *Pisaster* were recorded to be rare.

Menge (1979) also worked on two congeneric coexisting starfish *Asterias vulgaris* and *A. forbesi*. These showed great similarity in size, diet, time and place of occurrence with a positive correlation between their body size at different sites. Also in contrast to the previous pair no interference behaviour was observed, but the food records suggest that *A. forbesi* eats members of its own species more frequently than members of *A. vulgaris*; if this is substantiated it could contribute stability.

4.7 Lake-dwelling triclads

Reynoldson and his co-workers have extensively investigated European populations of *Polycelis nigra*, *P. tenuis*, *Dugesia lugubris* and *Dendrocoelum lacteum* (*see* Reynoldson, 1966, for a review) which coexist in productive lakes.

Field evidence for competition between them rests very heavily on demonstrations of food shortage and the overlap of food which they normally take. Their feeding similarities and differences were described (Reynoldson and Young, 1963) by examining gut contents of freshly collected animals and by giving them a choice, in the laboratory, of prey selected from the three important invertebrate phyla: annelids, arthropods and molluscs. *Dugesia* was unique in feeding on gastropods and *Dendrocoelum* was more able to capture active intact prey, but the overlap was otherwise large (Table 4.6).

Table 4.6 Broad classification of triclad food based on identification of skeletal remains from gut contents of animals collected in the field (From Reynoldson and Young, 1963).

| Triclad species | Percentage of prey group | | | |
	Oligochaeta	Arthropoda	Mollusca	Others
Polycelis nigra	68.6	25.7	0	5.7
P. tenuis	80.8	16.7	0	2.6
Dugesia lugubris	53.3	3.3	42.2	1.1
Dendrocoelum lacteum	88.8	8.9	0	2.2

They do not prey on each other and natural predation upon them is said to be rare. They show one unusual and important feature – they degrow when starved. Mature animals resorb gonads and the whole body shrinks at the rate of 1 mm per month (in *Polycelis*), shortening until they are confused with newly recruited young in a natural population. Death from starvation can therefore be postponed and the population fitted, with little lag, to the food supply.

Two types of test were performed in the field. Firstly, additional food (chopped earthworm) was shown to be followed by a spectacular increase and increased proportion of sexually mature animals (Reynoldson, 1964) (Table 4.7). Secondly, *Polycelis tenuis* was depleted by collection from a pond

Table 4.7 Effect of increasing food supply on the percentage increase of population size and the proportion of mature triclads in the populations (From Reynoldson, 1964.)

| | Total recruitment (%) | | Proportion of mature triclads (%) | |
	Control	Increased food	Control	Increased food
Triclad species				
Polycelis tenuis (College pond)	50	350	4.3	55.2
P. tenuis (New pond)	50	175	4.3	13.5
Dugesia lugubris	38	137	2.2	32.9
Polycelis nigra	90	500	–	–

(Reynoldson, 1966) and then allowed to recover. The ratio of *P. tenuis* to *P. nigra* was reduced to a tenth of its previous abundance (Table 4.8), but when removal ceased *P. tenuis* recovered in four years to its initial relative abundance. The absolute densities of the flatworms are not known, but this experiment provides strong evidence for stable coexistence.

Table 4.8 Changes of proportions of triclad species with removal of *Polycelis tenuis* in 1948–52 and subsequent recovery (From Reynoldson, 1966, courtesy Academic Press Inc. (London) Ltd.)

	1948	1952	1954	1955	Early 1956	Mid 1956	Late 1956	1957
Sample size . . .	500+	400	630	440	400	420	720	400
% *P. tenuis*	63	10	17	30	41	57	70	55
% *P. nigra*	13	23	35	38	28	15	14	15
% *Dugesia lugubris*	25	67	47	32	31	28	16	30

Most of these planarian species appear to fit Reynoldson's 'food refuge' conceptual model where each is superior to all the others at exploiting a different special food although there is considerable overlap in food range. Only the two *Polycelis* species show general similarity in food taken. In a more general context the word 'refuge' is far from ideal because each species needs not have some exclusive food, but merely be superior in exploiting it.

Sigurjonsdottir and Reynoldson (1977) report laboratory experiments in which 10 flatworms were used to start each culture, but various ratios of each species were used to test effects on each others' growth rates. The final ratios were compared with the initial ones after 12 weeks of experimental feeding regime. There was an overall tendency for average individual growth to be lower in monospecific cultures compared with those where some of the 10 individuals were replaced by individuals of different species. This should indicate stable coexistence and a graphical method of investigating return to equilibrium ratios was employed (*see* de Wit, 1961). The logarithm of the starting ratio of the pair of species is plotted against the logarithm of their ratio at the end of the experiment for each replicate. The results of this operation are self evident since, in a stable pairing, any input ratio above the equilibrium mixture should give a lower output ratio or vice versa. The points will lie along a line with a slope of less than 45° and the stable ratio is where this line crosses the 45° line. Other cases are unstable within the limits of the experiments.

Some experimental pairs confirmed prediction of coexistence when food variety was available for example *Dendrocoelum* vs. *Polycelis* with both *Asellus* and *Lumbricillus* as food (Fig. 4.7), these also coexisted with only *Lumbricillus* as food, but *Dendrocoelum* displaced *Polycelis* with only *Asellus* which was *Dendrocoelum*'s food refuge.

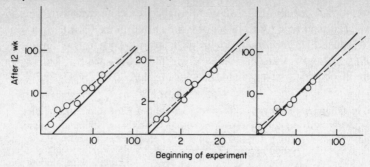

Figure 4.7 The ratio of biomass of the triclads *Dendrocoelum* and *Polycelis* at the start of each competition experiment (abscissa) plotted against the ratio of biomass after 12 wk of three different feeding regimes. Left, *Asellus* is the only food and any stable ratio is unrealistically extreme for the numbers supportable by these conditions; centre, *Lumbricillus* is the only food and there is a possible stable ratio of approx. 10 *Dendrocoelum*:1 *Polycelis*; right, both prey species are used and a stable ratio of 5 *Dendrocoelum*:1 *Polycelis* is predicted. (From Sigurjonsdottir & Reynoldson, 1977.)

4.8 The *Daphnia carinata* complex

Coexisting competitors may show extreme similarity and the work of Hebert (1977) provides such an example. *D. carinata* has been subdivided into nine species which are readily distinguishable by isozyme analysis, but even so they are genetically more similar, according to this technique, than members of other sibling groups such as the *Drosophila willistoni* group. Whatever the rank of these taxa they are interesting if they compete.

Three or four of the nine species frequently coexist in small Australian ponds. Most identification was done by morphology after establishment of the appropriate limits by isozyme analysis, but this continued to be used in difficult cases.

Fecundity is used to test for competitive effects. Egg number is related to size therefore a regression of the logarithm of egg number on body length was calculated for each of the 101 samples and used to estimate the egg production for a standard 2.5 mm female. This was multiplied by the ratio of egg producers to non-producers in the population to give a value called standard egg production (SEP) – a measure of reproductive rate independent of the size structure of the population. SEP was inversely correlated with density of *Daphnia* indicating reduced fecundity as the result of competition. Where *D. carinata* and *D. cephalata* were together there was no case where one had a high SEP and the other a low one. High total density correlated with low SEP in both and SEP tended to show an overall positive correlation between the two species (Table 4.9). It was noted that *D. carinata* reproduced better in winter while *D. cephalata* was better in summer, but temporal trends were similar in the two species (Table 4.10).

Table 4.9 Densities of *Daphnia carinata* and *D. cephalata* in populations where standard egg production was very low (<1.0 in both spp.) and combined density high (>3000 adults/500 1) (From Hebert, 1977.)

D. carinata	D. cephalata	Combined
22961	4	22965
5828	2	5830
3494	873	4367
1702	1846	3548
866	2598	3464
7	3168	3175

Hebert concluded that the same resources were being used and this caused parallel changes in SEP so that interspecies competition was closely equivalent to intraspecies competition in these two species. Other species pairs were more similar morphologically and may also show this sort of ecological equivalence.

Table 4.10 Standard egg production (S.E.P.) in subsequent samples of coexisting *Daphnia* at one site (From Hebert, 1977.)

S.E.P.		Change in S.E.P.	
D. carinata	D. cephalata	D. carinata	D. cephalata
18.52	22.83		
1.84	3.19	−	−
1.33	0.0	−	−
15.13	10.28	+	+
0.09	1.19	−	−
4.10	9.62	+	+
3.36	0.72	−	−
1.56	0.05	−	−
1.10	0.0	−	−

This example is particularly important in view of the unrealistic emphasis which has been laid on similarity as a cause of displacement. The evidence for competition is indirect, however, and experimental manipulation would be more convincing.

No interference behaviour was recorded and there seems to be no reason why a filter-feeding animal should specialize on part of the food spectrum available. The same work has to be done whether all the particles of food are used or not and it could be argued that the extra work of selection is a positive disadvantage. If the species are effectively the same then long-term coexistence is dependent on chance (Pontin, 1963) and shows neutral stability. Subtle differences giving positive stability cannot be discounted.

4.9   Leafhoppers

These Homoptera include a number of examples of genera which possess many species on the same hostplant. Ross (1957) studied six species of *Erythroneura* which breed on *Platanus occidentalis* in eastern USA. They have closely similar morphology and only two are readily distinguishable by colour pattern. Males are distinguishable in all known cases by their genitalia.

Their distributions vary in extent, but all six live together in Illinois where they were studied, and they also show differences in abundance with *E. bella*, usually rare. *E. lawsoni* showed the widest tolerance of humidity, extending further into dry areas, but also occurring with others in sheltered humid sites. No other consistent ecological differences were noted. Parasitism by dryinid Hymenoptera was recorded, but not investigated as a cause of coexistence. Species diversity was higher at higher densities – the opposite of competitive exclusion.

McClure and Price (1975) followed up this work and included two further species, but at their study areas a maximum of seven were found together. To test for interactions enclosures were made. Single *Platanus* leaves were caged *in situ* and various starting densities of single species or species pairs were used. The densities, per leaf, chosen included those found by sampling and four generations per year were possible.

In the enclosures progeny were reduced in number by 70% per female if six females were introduced instead of one to start the breeding populations. No progeny matured at all where 14 females were introduced and some reduction was recorded if only two were used – another example of the progressive effect rather than threshold effect of competition.

Individuals of the same species were shown to compete more severely than individuals of different species (Fig. 4.8) since a larger total of progeny was produced by the same number of females in a mixture of two species although not all combinations showed a statistically significant increase. Dominance of one species over another was not shown, but *E. lawsoni* was more successful at high densities.

Feeding damage of up to 98% of the leaf area was found in late summer and there can be little doubt that the species are competing. The cause of coexistence is presumably the higher severity of intraspecific competition and there seems no need to look further for another mechanism of coexistence although McClure and Price do.

Interference is not recorded, but parasitism by the dryinid *Aphelopus* was found to reach 9.8% for all hosts combined. Interestingly it could distinguish these very similar *Erythroneura* species and the most abundant host *E. arta* was parasitized much more frequently than chance. It suffered 26% parasitism – thirteen times the frequency of the next most abundant host. McClure and Price state that the per cent parasitism stays constant with increasing density of hosts and is therefore not density dependent, but they make no

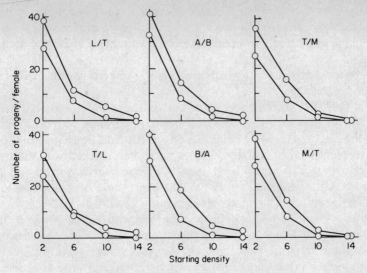

Figure 4.8 Mean number of progeny produced per female (ordinate) at four starting densities (abscissa) for various species of *Erythroneura* leafhoppers. L/T, number of progeny per female *E. lawsoni* in unispecific groups for the lower curve and in the same-sized groups composed of equal starting numbers of it and *E. torella* for the upper curve; T/L, similarly for pure *E. torella* in the lower curve and with *E. lawsoni* in the upper curve. For the other plots: A, *E. arta*; B, *E. bella*; and M, *E. morgani*. (*From* McClure and Price, 1975.)

comment on the possibility of apostatic effect of the parasite in stabilizing coexistence.

It would be most interesting to know the underlying cause for intraspecific competition exceeding interspecific competition in these animals.

4.10 Other enclosure experiments

Although enclosures introduce artificial conditions which may make experiments less conclusive they do have two valuable functions. As well as providing the opportunity to start experiments with known populations they can also be used to exclude potential competitors from an otherwise natural population.

Haven (1973) worked on two limpet species *Acmaea scabra* and *A. digitalis*. Fences, 4.5 cm high, of stainless-steel mesh were epoxied to rock surfaces to enclose plots of 20 by 40 cm and were used to measure the growth rates of each species after removal of the other. Individuals grew faster compared with controls and this was most marked in *A. scabra* after removal of *A. digitalis*. The algal crop also increased and therefore was presumably restricting growth before the experiments. There was no evidence of interference or differential exploitation, but there was some evidence of *A. digitalis* affecting the growth rate of members of its own species more than that of *A. scabra*.

Longer-term experiments which show change of density or fecundity would be more convincing.

Istock (1973) studied 10 species of corixid (false-waterboatmen) breeding in one Michigan pond and used enclosures of nylon cloth with 0.25-mm mesh to test the effects of rearing combinations of two and three species together. They were emptied of corixids in spring and then stocked with known high densities of the two commonest species, *Hesperocorixa lobata* and *Sigara macropala*, either together or alone. The cages were open topped and this allowed invasion from the ambient populations.

H. lobata reached a higher density alone than with *Sigara*, but no reciprocal effect was found. Istock suggested that the other rarer species coexist with either of the common two because chance plays some part in exploitation of detritus by these insects and 'errors of exploitation' allow the rarer species to 'coexist'. Long-term stability is not to be expected from this behaviour alone.

4.11 Coexistence in plants

Although most habitats are occupied by several species of photosynthetic plants there are very few examples where anything is known about the causes of coexistence. Interference competition for space and identical food requirements would be expected to unstabilize coexistence so research is certainly needed to distinguish between examples of true stability and spurious cases of slow displacement or underlying microhabitat mosaic.

The poppies investigated by Harper and McNaughton are therefore of particular interest (see Harper, 1961; Harper and McNaughton, 1962). Five species of *Papaver* were found to occur together in southern Britain and the commonest, *P. rhoeas*, frequently coexisted with the others. The others were not found alone. *P. dubium* and *P. lecoqii* are very similar to *P. rhoeas*, but can be distinguished by chromosome number or latex colour and these three are selected here for simplicity.

Seeds were sown in garden plots in combinations of species pairs, but they produce a high proportion of dormant seeds so the numbers quoted refer to those expected to germinate after sowing. The chance of a seed producing a plant is much higher (Fig. 4.9) at low density, as expected, and also if the other seeds are predominantly the opposite species except for the pair *P. dubium* and *P. lecoqii* where seeds of either species have equivalent effect.

Self-thinning makes each species largely independent of the other in spite of the severity of crowding which causes more than 50% loss at all the experimental densities. The actual mechanism was not discovered and it

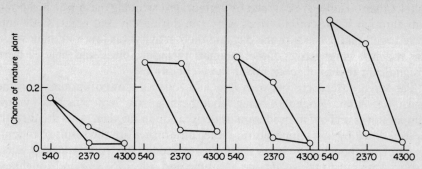

Figure 4.9 The chance of a poppy seed producing a mature plant (ordinate) in three densities of mixed stands. In each case the low density is started by 270 seeds of each species of the pair, the intermediate density by 270 of one species and 2100 of the other while the high density has 2100 of each. The graphs from left to right are: *Papaver dubium* in the presence of *P. rhoeas*; *P. rhoeas* in the presence of *P. dubium*; *P. lecoqii* in the presence of *P. rhoeas*; and *P. rhoeas* in the presence of *P. lecoqii*. The difference between the two intermediate densities measures the difference between 'self-thinning' and thinning by alien species. In all these examples intraspecific competition exceeds interspecies competition in severity. (*From* Harper, 1961.)

would be very interesting to know exactly how one plant selectively inhibits members of its own species more than those of different species. Stable coexistence would of course be expected to result and some interspecific competition would be recurrent.

Other cases of auto-inhibition are known and the example of the rain-forest tree *Grevillea* which kills its own seedlings (Webb, Tracy and Haydock, 1967) may give increased diversity of species. *Typha latifolia* the reedmace inhibits germination of its own seeds (McNaughton, 1968), but this plant normally establishes pure stands by vegetative spread.

Whatever the mechanism used one cannot expect natural selection to favour intraspecific killing and consequent increased survival of competing species so auto-inhibition either has some other function or is the inevitable result of intraspecific competition.

4.12 Distantly related competitors

Animals could hardly be more dissimilar than ants and rodents and still be in the same trophic level of a community. However, seed-eating ants obviously exploit an overlapping range of food with rodents and studies of Arizona desert species have shown (Brown, Grover, Davidson and Lieberman, 1975; Brown and Davidson, 1977) that they do compete.

Seeds of several size classes and species were presented as a mixture in ashtrays and re-examined after desert animals had been given an opportunity to take them. Half were covered by vertebrate-excluding mesh which allowed ants through, and separate trials were run during day and night. Only ants were found taking seeds in the day and their activity was lower at night when the rodents were active. Other animals including birds and beetles were insignificant takers of seeds under these conditions.

The large differences between ants and rodents showed up mainly in the diurnal rhythm. Rodents, being endothermic, are less affected by low temperatures and avoid predation or desiccation in the day. Ants are directly slowed down by low temperature. Other differences are the result of rodents using cheek pouches to harvest clumps of seeds effectively and less selectively. Ants must take seeds one at a time and select for size and roughness which allows the mandibles enough grip. As a result rodents can achieve 70–80% removal of seeds from ashtrays while ants remove 20% at the most in the 12 h before renewal.

There was a wide range of overlap in seeds taken and the difference in activity rhythm is irrelevant because naturally provided seeds are renewed at much longer intervals. The rodent species concerned were mainly *Dipodomys spectabilis*, *D. merriami*, *Perognathus penicillatus*, *P. baileyi*, *P. flavus*, *Peromyscus eremicus* and *P. maniculatus*. The ant species included *Pogonomyrmex rugosus*, *P. desertorum*, *Novomessor cockerelli*, *Solenopsis xyloni* and several *Pheidole* species.

To test the effects of the potential competitors on each other, exclusion experiments were performed. Circular plots of 36 m diameter with three treatments plus controls were set up; two plots with rodent-excluding mesh fences and the rodents trapped to extinction, two plots treated with insecticide to remove ants and thirdly two plots with both ants and rodents removed. The first treatment showed an increase of 71% in ant colonies, the second a 24% increase in rodent biomass and the third a five times greater seed density after three years (Table 4.11). Presumably the ant increase is real, but some measurement of colony size is strictly necessary to be sure.

Dipodomys merriami and the *Pheidole* species showed the largest increases and these are the species observed to be most effective at harvesting dense accumulations of seeds so might be expected to increase in the altered environment.

Table 4.11   The results of excluding ants or rodents from plots of desert in which they would both normally eat seeds (From Brown and Davidson, 1977.)

	Rodents excluded	Ants excluded	Control	Increase (%) relative to control
Ant colonies	543	—	318	71
Rodent no.	—	144	122	18
biomass	—	5.12 kg	4.13 kg	24

Coexistence is likely to have lasted for millions of years, but natural selection has not eliminated competition and it should not be expected to do so. There is obviously a strong counterselection against specializing on some seeds with no resulting advantage in reduced searching effort.

4.13   The effects of herbivores and predators

The outcome of competition can be influenced by many other factors especially, of course, if they act differentially on the competing species. The factor need not be discriminatory as noticed by Darwin (1859) who described the effect of ceasing to mow short turf. Twenty species were present while it was kept short, but nine of these were lost if the turf was allowed to grow long. Presumably some small plants can survive in competition with grass unless the grass can overshadow them. Slobodkin (1964) showed that *Chlorohydra* displaced *Hydra* in laboratory culture if kept in the light, but removing a fixed percentage of each at regular intervals allowed persistence of both species. 90% artificial 'predation' was needed to keep them together for over 200 days with no sign of continuous decline of either species. The explanation is probably that repeated lowering of density reduces the incidence of interference to a level which does not override the stabilizing effects of differences in exploitation. They coexist in the dark without predation.

Herbivores and predators are usually more selective than this and the method used for grazing by small mammals or birds compared with large grazers is important to the composition of grassland. The effect of rabbits (*Oryctolagus cuniculus*) was well documented by Tansley and Adamson (1925) who excluded rabbits from plots of chalk grassland. The wet weight of vegetation inside the enclosures was doubled and its height increased 3–5 times that outside. The diversity inside was much lower and only one dicotyledon species, *Poterium sanguisorba*, reached more than 5% cover whereas four species outside reached more than 5%. Short rabbit-resistant species declined markedly where rabbits were removed, for example *Cirsium acaule* decreased from 14% cover to nil.

The percentage cover of grasses increased from 41.6 to 67.2, but the dominant species was different. *Festuca ovina* was the most abundant plant under rabbit grazing, but *F. rubra* took over in the enclosures and *Avena pratensis* also increased. These may well have been preferred food of the rabbits which clearly select individual leaves when feeding. The effects of rabbits are not simply from grazing, but a mosaic of local habitat differences is produced by soil disturbance near burrow systems and they tend to use concentrated social latrines. Patches of toxic plants such as *Senecio jacobaea*, *Urtica*, *Solanum* and *Sambucus* are typical of areas where rabbits are concentrated.

Other patches in this habitat are produced by other animals locally altering the results of competition. The patches of *Thymus* and other plants typical of *Lasius flavus* anthills (*see* King, 1977), not molehills, were also recorded by Tansley and Adamson.

The overall effects of rabbit grazing were spectacularly confirmed by the introduction of myxomatosis into Britain in 1953. Species which had previously been eliminated by rabbits colonized grassland and this was particularly noticeable in the case of shrubs and trees. Rabbits had been having a strong delaying effect on the competitive displacement producing succession to woodland (*see* Sheail, 1971).

Animal grazing is further reviewed by Harper (1977).

Connell (1971) suggests that the heterogeneity of tree species in tropical rain forest is the result of herbivore attack. The mortality of saplings was shown to be higher if next to a mature tree of the same species compared with mature trees of different species (Table 4.12) in an area of North Queensland.

Table 4.12 Mortality of tree saplings in rain forest when next to conspecific neighbours compared with mortality next to different species (From Connell, 1970.)

| | % mortality | |
| | | |
Height (m)	Conspecific neighbours 376 in 1965, 306 in 1969	Neighbours of different sp. 1735 in 1965, 1605 in 1969
0.11–0.19	37.5	18.2
0.2–0.35	21.6	11.2
0.36–0.79	11.0	6.6
0.8–6.34	3.4	3.0

Three explanations were offered: autotoxic effects, nutrient specialization giving more severe intra- than interspecific competition, and thirdly a permanent population of specific herbivores on the mature trees which eliminates saplings of the same species locally. Trenching around plots to prevent root competition did not improve the relative survival of seedlings of the same species as adjacent mature trees.

The earlier work of Connell (1961*a*, *b*, 1971) on predation of barnacles by the gastropod *Thais* (*Nucella*) used experimental exclusion of *Thais* from areas caged by stainless-steel mesh. Reduced competition between *Balanus balanoides* and *Chthalamus stellatus* at lower shore levels was produced by the tendency of *Thais* to eat larger barnacles thus reducing the effect of larger and quicker growing *Balanus*. However, much competitive displacement has already occurred in natural populations before the barnacles are large enough to be attractive to *Thais*. Another barnacle/*Thais* interaction was investigated by Connell (1971) on San Juan Island, USA. Here two species of *Balanus* formed zones on the shore, but in this case the zonation was caused by preferential attack on *B. glandula* by predators confined to the lower shore.

The lower limit of the barnacle zone in New England was found by Menge (1976) to be caused by competition for space with mussels *Mytilus edulis*, but *Thais* was important in less-exposed situations. Displacement by mussels is a far-reaching factor in Pacific coast communities as Paine (1966) showed by removing *Pisaster* which was an important predator of mussels. *Mytilus* increased enormously to dominate much of the shore in the absence of *Pisaster* and the community was reduced from 15 to 8 species.

Introduction of fish into ponds and lakes has been followed by a reduction in the larger zooplankton (e.g. Hrabacek, Korinek and Prochazkova, 1961) and comparison of gut contents with plankton samples (*see* Green, 1967) has shown selective eating of larger prey. Dodson (1974) gives an example of the effect this behaviour may have on competition and shows the sort of complication which is possible in natural communities. *Daphnia minnehaha* coexists with the larger *D. middendorffiana* in plankton cages which exclude predators, but *Diaptomus shoshone* selectively eats the smaller prey and the smaller *D. minnehaha* is displaced. However, *Diaptomus* is purple and opaque so it is eaten selectively by visual predators such as salamanders. The two *Daphnia* species can thus coexist. Another predator, the transparent midge larva *Chaoborus*, withstands vertebrate predation and eats *Diaptomus*, but if vertebrates are absent *Diaptomus* gets the better of *Chaoborus* by eating eggs and small larvae. The consequences of all this to *Daphnia* competition are predictable only from a very wide range of knowledge of the community and are certainly not predictable from the overlap of food taken by the competitors themselves.

The most interesting predator behaviour with respect to competition (Williamson, 1957) is the tendancy to take a disproportionately large number of the most numerous species of prey. A predator may learn the appearance of the most abundant type of prey or the site where it is found simply because it meets with success most frequently in association with these stimuli. This has been called the formation of a searching image (Tinbergen, 1960) and the advantage of such learning is, of course, that the predator only does the work necessary to obtain a particular type of food if it is rewarded enough for the effort to be profitable. Holling (1959) working on predation by small mammals on sawfly cocoons with an alternative food supply showed that the proportion of sawfly cocoons eaten increased with the density provided until satiation countered the effect.

Schneider (1978) has shown a striking example in the field. More than five thousand shore birds were observed each day to exploit the invertebrates of Plymouth estuary Massachusetts during the period from mid-July to mid-August each year. Four species made up the bulk of the bird numbers and these were: the sanderling (*Calidrus alba*), the semipalmated sandpiper (*C. pusilla*), the short-billed dowitcher (*Limnodromus griseus*) and the black-bellied plover (*Pluvialis squaterola*). Each showed some differences in prey selection, but crop contents showed them to be taking the larger and commoner invertebrates including *Nereis*, *Tellina* and *Crangon* species rather than smaller species such as *Hydrobia*.

To estimate the effect of this predation on the invertebrate fauna 20 core samples were taken from each of 13 representative sites before and after the birds' visit. Wire-mesh cages were used to exclude large predators from comparable areas and make the investigation a direct test of cause and effect. The only other predators noted were flatfish and *Limulus*, but the flatfish were absent at this season while the *Limulus* distribution did not correspond to the results obtained.

The proportion lost of each of the main prey species corresponded to the initial abundance (Table 4.13). The most abundant species, except *Gemma*, showed the most substantial loss and the overall effect was an equalization of prey species numbers. *Gemma*, a thick-shelled clam less than 4 mm long, was presumably unpalatable and was not often found in crop contents. The numbers inside the enclosures did not decline and the same general results were obtained two years running.

Table 4.13 Mortality of intertidal invertebrates preyed on by shore-birds (From Schneider, 1978, reproduced here by courtesy of *Nature*.)

	Rank of species in order of abundance			
	1	2	3–5	>5
No. of species . . .	6	11	21	32
Average initial density per m^2	1158	331	335	179
Average mortality (%)	84	78	67	36

4.14 Conclusions

There are examples from a very wide range of organisms with evidence of competition although the competing species continue to coexist.

Stability from intraspecific competition exceeding interspecific competition is strongly supported for most of these examples. There is also another possible source of stability from predator behaviour and coexistence is certainly favoured by this in grassland herbs and shore animals.

The degree of similarity or taxonomic relationship of competitors varies from the near identical to the vastly different. Species are not necessarily the lowest rank for competitors and it is probable that clones also compete. Perhaps the great variety between coexisting clones of plants such as dandelion (*Taraxacum*) exists because of the stabilizing process shown by interspecific competition.

Stable competition is certainly the main reason why the share-out of energy in communities is so complex, but the competition may have been reduced by long coexistence and natural selection.

References

Betts, M. M. (1955). The food of titmice in oak woodland. *J. Anim. Ecol.* **24**, 282–323.

Bergström, G. and Löfqvist, J. (1970). Chemical basis for odour communication in four species of *Lasius* ants. *J. Insect Physiol.* **16**, 2353–75.

Brown, J. H., Grover, J. J., Davidson, D. W. and Lieberman, G. A. (1975). A preliminary study of seed predation in desert and montane habitats. *Ecology* **56**, 987–92.

Brown, J. H. and Davidson, D. W. (1977). Competition between seed eating rodents and ants in desert ecosystems. *Science* **196**, 880–2.

Connell, J. H. (1961a). The effects of competition, predation by *Thais lapillus*, and other factors on natural populations of the barnacle *Balanus balanoides*. *Ecol. Monogr.* **31**, 61–104.

Connell, J. H. (1961b). The influence of interspecific competition and other factors on the distribution of the barnacle *Chthalamus stellatus*. *Ecology* **42**, 710–23.

Connell, J. H. (1971). On the role of natural enemies in preventing competitive exclusion in some marine animals and in rain forest trees. In *Proc. Advan. Study Inst.*, 298–506, *Dynamics of Numbers in Populations*, Oosterbeek 1970, ed. P. J. den Boer and G. Gradwell. Wageningen: Centre for Agricultural Publishing and Documentation (PHDOC).

Crombie, A. C. (1945). On competition between different species of graminivorus insects. *Proc. Roy. Soc.* B **132**, 362–95.

Crombie, A. C. (1946). Further experiments on insect competition. *Proc. Roy Soc.* B **133**, 76–109.

Darwin, C. (1859). *The Origin of Species by Means of Natural Selection.* Harvard Facsimile 1st edition; 1964.

Dhondt, A. A. (1977). Interspecific competition between great and blue tit. *Nature* **268**, 521–3.

Dodson, S. I. (1974). Zooplankton competition and predation: an experimental test of the size-efficiency hypothesis. *Ecology* **55**, 605–13.

Donisthorpe, H. St J. K. (1927). *British Ants their Life-History and Classification.* London: Routledge.

Gibb, J. (1954). Feeding ecology of tits, with notes on trecreeper and goldcrest. *Ibis* **96**, 513–43.

Green, J. (1967). The distribution and variation of *Daphnia lumholtzi* (Crustacea: Cladocera) in relation to fish predation in Lake Albert, East Africa. *J. Zool.* **151**, 181–97.

Harper, J. L. (1961). Approaches to the study of plant competition. *Symp. Soc. Exp. Biol.* **15**, 1–39.

Harper, J. L. (1977). *Population Biology of Plants.* London, New York: Academic Press.

Harper, J. L. and McNaughton, I. H. (1962). The comparative biology of closely related species living in the same area. VII. Interference between individuals in pure and mixed populations of *Papaver* species. *New Phytol.* **61**, 175–88.

Hartley, P. H. T. (1953). An ecological study of the feeding habits of the English titmice. *J. Anim. Ecol.* **22**, 261–88.

Haven, S. B. (1973). Competition for food between the intertidal gastropods *Acmaea scabra* and *Acmaea digitalis*. *Ecology* **54**, 143–51.

Hebert, P. D. N. (1977). Niche overlap among species in the *Daphnia carinata* complex. *J. Anim. Ecol.* **46**, 399–409.

Högstedt, G. (1980). Prediction and test of the effects of interspecific competition. *Nature* **283**, 64–66.

Holling, C. S. (1959). The components of predation as revealed by a study of small-mammal predation of the European pine sawfly. *Can. Entomol.* **91**, 293–320.

Hrabacek, J. M., Korinek, D. V. and Prochazkova, L. (1961). Demonstration of the effect of the fish stock on the species composition of zooplankton and the intensity of metabolism of the whole plankton association. *Verh. Internat. Verein. Limnol.* **14**, 192–5.

Istock, C. A. (1973). Population characteristics of a species ensemble of waterboatmen (Corixidae). *Ecology* **54**, 145–59.

Jackson, J. B. C. and Buss, L. (1975). Allelopathy and spatial competition among coral reef invertebrates. *Proc. Nat. Acad. Sci. USA* **72**, 5160–3.

King, T. L. (1977). The plant ecology of ant-hills in calcareous grasslands. I. Patterns of species in relation to ant-hills in southern England. *J. Ecol.* **65**, 235–56.

Krebs, C. J. (1972). *Ecology: the Experimental Analysis of Distribution and Abundance*. Harper and Row.

McClure, M. S. and Price, P. W. (1975). Competition among sympatric *Erythroneura* leafhoppers (Homoptera: Cicadellidae) on American sycamore. *Ecology* **56**, 1388–97, (The Ecological Society of America).

McNaughton, S. J. (1968). Autotoxic feedback in relation to germination and seedling growth in *Typha latifolia*. *Ecology* **49**, 367–9.

Menge, B. A. (1972). Competition for food between two intertidal starfish species and its effect on body size and feeding. *Ecology* **53**, 635–44, (The Ecological Society of America).

Menge, B. A. (1976). Organisation of New England rocky intertidal community: role of predation, competition, and environmental heterogeneity. *Ecol. Monogr.* **46**, 355–93.

Menge, B. A. (1979). Coexistence between the seastars *Asterias vulgaris* and *A. forbesi* in a heterogeneous environment: a non-equilibrium explanation. *Oecologia* **41**, 245–72.

Menge, J. L. and Menge, B. A. (1974). Role of resource allocation, aggression and spatial heterogeneity in coexistence of two competing intertidal starfish. *Ecol. Monogr.* **44**, 189–209.

Odum, E. P. and Pontin, A. J. (1961). Population density of the underground ant, *Lasius flavus*, as determined by tagging with P^{32}. *Ecology* **42**, 186–8.

Paine, R. T. (1966). Food web complexity and species diversity. *Am. Naturalist* **100**, 65–75.

Perrins, C. M. (1979). *British Tits*. London: Collins.

Pontin, A. J. (1960). Field experiments on colony foundation by *Lasius niger* (L.) and *L. flavus* (F.) (Hym., Formicidae). *Insectes Sociaux* **7**, 227–30.

Pontin, A. J. (1961a). Population stabilization and competition between the ants *Lasius flavus* (F.) and *L. niger* (L.). *J. Anim. Ecol.* **30**, 47–54.

Pontin, A. J. (1961b). The prey of *Lasius niger* (L.) and *L. flavus* (F.) (Hym., Formicidae). *Ent. Mon. Mag.* **97**, 135–7.

Pontin, A. J. (1961c). Observations on the keeping of aphid eggs by ants of the genus *Lasius* (Hym., Formicidae). *Ent. Mon. Mag.* **96**, 198–9.

Pontin, A. J. (1963). Further considerations of competition and the ecology of the ants *Lasius flavus* (F.) and *L. niger* (L.) *J. Anim. Ecol.* **32**, 565–74.

Pontin, A. J. (1969). Experimental transplantation of nest-mounds of the ant *Lasius flavus* (F.) in a habitat containing also *L. niger* (L.) and *Myrmica scabrinodis* Nyl. *J. Anim. Ecol.* **38**, 747–54.

Pontin, A. J. (1978). The numbers and distribution of subterranean aphids and their exploitation by the ant *Lasius flavus* (Fabr.). *Ecol. Entom.* **3**, 203–7.

Reynoldson, T. B. (1964). Evidence for interspecific competition in field populations of triclads. *J. Anim. Ecol.* **33**, 187–201.

Reynoldson, T. B. (1966). The distribution and abundance of lake-dwelling triclads – towards a hypothesis. *Adv. Ecol. Res.* **3**, 1–71.

Reynoldson, T. B. and Young, J. O. (1963). The food of four species of lake-dwelling triclads. *J. Anim. Ecol.* **32**, 175–91.

Ross, H. H. (1957). Principles of natural coexistence indicated by leafhopper populations. *Evolution* **11**, 113–29.

Schneider, D. (1978). Equalisation of prey numbers by migratory shorebirds. *Nature* **271**, 353–4.

Sheail, J. (1971). *Rabbits and their history*. Newton Abbot: David and Charles.

Sigurjonsdottir, H. and Reynoldson, T. B. (1977). An experimental study of competition between triclad species (Turbellaria) using the de Wit model. *Act. Zool. Fennica* **154**, 89–104.

Skellam, J. G. (1951). Random dispersal in theoretical populations. *Biometrika* **38**, 196–218.

Slobodkin, L. B. (1964). Experimental populations of Hydrida. *J. Anim. Ecol.* **33**, 131–48.

Tansley, A. G. and Adamson, R. S. (1925). Studies of the vegetation of the English chalk. III. The chalk grasslands of the Hampshire–Sussex border. *J. Ecol.* **13**, 177–223.

Tinbergen, L. (1960). The natural control of insects in pinewoods: I. Factors influencing the intensity of predation by song birds. *Arch. Néerl. Zool.* **13**, 265–343.

Webb, L. J., Tracey, J. G. and Haydock, K. P. (1967). A factor toxic to seedlings of the same species associated with living roots of the non-gregarious subtropical rain forest tree *Grevillea robusta*. *J. Appl. Ecol.* **4**, 13–25.

Williamson, M. H. (1957). An elementary theory of interspecific competition. *Nature* **180**, 422–5.

de Wit, C. T. (1961). Space relationships within populations of one or more species. *Symp. Soc. Exp. Biol.* **15**, 314–29.

5 Displacement

This means exclusion of one species by another from a habitat which it would otherwise occupy. Experimental test of this is as important as testing coexisting species for competition as exclusion may well result from other causes, such as predation, which are outside the scope of this book.

5.1 The conditions for displacement

In general terms there are three types of situation which could result in one species causing another in the same trophic level to become extinct.

(1) Interference competition may override the stabilizing effects of exploitive competition. A species which becomes dominant because of qualitative differences in its aggressive methods or because it is larger and more powerful in combat may drive out or directly exterminate its competitor. The main exceptions will be found in mobile species where the subordinates have evolved avoidance behaviour or their smaller size enables them to take refuge in spaces too small for the dominant to enter. It is perhaps worth re-emphasizing that the main difference between the action of exploitation and interference results from individuals of the same species being evenly matched in combat compared with those of different species and in the case of interference this is of course destabilizing.

(2) Species exploiting common resources may differ in efficiency and resulting growth rate so much that the only possible equilibrium mixture gives a density which is too low for one species to maintain itself against other factors or to find mates. Species which are distantly related in the taxonomic sense and which adopt widely differing strategies may provide more examples, but we can only speculate about this because research has been biased towards closely related species.

(3) If the environment is effectively uniform then it could be argued that competitors cannot show their potential differences in exploitation. Displacement would then occur by chance in a manner analogous to genetic drift and accidents of colonization could predetermine the winner. The flour beetles *Tribolium confusum* and *T. castaneum* in laboratory culture (e.g. Park, 1954) may be an example where culture methods prevent differential exploitation. The suggestion that diversity creates stability (*see* Elton, 1958) has been

criticized on theoretical grounds (May, 1973), but its corollary – uniformity creates instability – should not be dismissed completely. It may be rare for natural environments to be uniform enough to cause instability and even laboratory cultures have edges and gradients of depth, but we know so little about reality in this area that it is important not to prejudge the issue.

Another cause of displacement is often considered. If one species exploits only some of the range of resources of another, but has no resources which are not exploited by the other then it could be displaced – it has no 'food refuge'. However, specialization must have advantages or it would not have been evolved and it presumably makes the specialist more likely to use its own resources than the generalist. *Lasius flavus* is in this position with respect to *L. niger*, but is much better adapted to its most important resources – subterranean aphids – than is the generalist *L. niger*. Perhaps species which are geographically separated could show displacement by this sort of superior exploitation if they are matched as the result of introduction, but unambiguous examples are not available (Chapter 7.4).

5.2 Sessile animals

There are very few examples where absence of one species of animal has been demonstrated to be caused by the presence of another species. Most probable instances of displacement have not been backed by experimental evidence, presumably because the observer thought it unnecessary to prove something expected from a general 'law of competitive exclusion'. The much quoted example provided by Connell (1961) working on intertidal barnacles will therefore be used again.

Larvae of *Balanus balanoides* and *Chthalamus stellatus* settle over a wide range of tidal level around the British coast, but *Balanus* grows quicker and larger on the lower shore where it displaces *Chthalamus* by growing over it or undercutting it. This was elegantly demonstrated by transplanting rocks down the shore from the *Chthalamus* zone, where *Balanus* does not survive the exposure, to the *Balanus* zone and bolting them in place. Half the area of each transplated rock was left as a control and the other half had *Balanus* removed by a needle where they touched or surrounded *Chthalamus*. Loss or covering-over of *Chthalamus* was mapped and the survivorship of this species was strikingly less in most cases where *Balanus* was left. *Chthalamus* survived well at lower levels when protected from *Balanus*, but no intraspecific crowding was noted in *Chthalamus* during the year's duration of the experiments. Intraspecific crowding was sometimes severe in *Balanus*, but individuals of the same species were presumably more evenly matched compared with the interaction between taller, quicker-growing individuals of *Balanus* and smaller *Chthalamus*.

Menge (1976) tested the effect of mussel (*Mytilus edulis*) removal on

Balanus in New England. Barnacles persisted much longer if the mussels were removed where they were dominant in mid-intertidal sites and this could determine the lower limit of *Balanus*. Competition for space using overgrowth is again the conspicuous interference behaviour giving displacement.

There is a possible parallel example among ants. In S. Britain *Myrmica ruginodis* is characteristic of woodland with some gaps in the canopy and it tolerates shade better than other *Myrmica* species. It does prefer warm sites and in habitats less suitable for the other species including sphagnum bogs or higher altitudes it may be found in the full sun. Brian (1952) found that *M. scabrinodis*, a sunny-site species, was dominant to *M. ruginodis* and displaced it by direct aggression from a sunny site. Plant succession could be partitioned by these ants with *M. rubra* as a possible intermediate species.

5.3   Mobile animals

Exclusion of mobile animals is difficult to demonstrate and the problems are well illustrated by Davis (1973) working in California on winter flocks of Juncos (*Junco hyemalis*) and golden-crowned sparrows (*Zonotrichia atricapilla*). The larger golden-crown occupied part of the study area with thick cover of *Rhamnus* and *Salix* while the junco was found in the remaining open part of the area. There were also differences in food as the juncos were primarily seed eaters and the golden-crowns ate mainly sprouting annuals.

A trapping programme was undertaken and 423 golden-crowns were ringed compared with 838 juncos using 50 traps over 4 years. Juncos were experimentally removed as a first attempt to test for any interaction and merely resulted in replacement by immigrants from the surrounding area. It was found impracticable to reduce junco numbers to any significant extent. The alternative removal of golden-crowns was successful and the junco population spread into the area previously occupied by golden-crowns (Fig. 5.1). Release of the captured birds resulted in a return to normal with immediate occupation of their preferred habitat. Only one direct interspecific attack was recorded – a golden-crown displacing a junco – but it is clear that the 55% heavier golden-crown, although outnumbered 2:1, effectively reduced the numbers of juncos in dense cover. Intraspecific aggression was much more frequent and it seems likely that the smaller bird normally avoided the larger species.

Balat (1977) had difficulty in removing *Passer montanus* from Moravian study areas because it was rapidly replaced in the same manner as Davis' juncos, but after three times the normal population was removed an increase in the *Parus major* numbers using nestboxes was recorded. Direct aggression is presumably the main cause and the sparrows recovered if allowed.

Similar experiments have been performed on small mammals. For example Redfield, Krebs and Taitt (1977) working near Vancouver used three

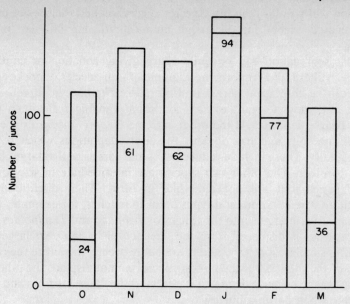

Figure 5.1   Estimates of the total population number of juncos and the percentage of them in the part of the study area occupied by *Salix* thicket from October to March, inclusive. Golden-crowned sparrows preferred this thicket and were removed from the beginning of November to the beginning of February. (*From* Davis, 1973.)

grassland areas where *Microtus* were normally well established, but *Peromyscus maniculatus* was normally rare in this habitat. *Peromyscus* was present in forested areas at a density of around 30 mice per ha at the end of the breeding season. Removal of *Microtus* from one grassland area gave a big increase in *Peromyscus* compared with control and the numbers reached 61 per ha after two years. Reinvasion by *Microtus* reversed this and *Peromyscus* was obviously excluded from grassland by *Microtus* under normal conditions.

The third area was particularly interesting because *Peromyscus* increased (34/ha) here in the presence of an abnormal *Microtus* population. Density of female *Microtus* was kept low here for another study and it may indicate that aggression comes mainly from female *Microtus* or that the social structure was disorganized enough for normal aggression toward *Peromyscus* to be reduced.

No food shortage was noticed so any advantage to *Microtus* is more subtle or displacement is an accidental consequence of *Microtus* behaviour.

Stoecker (1972) also worked on a species pair, *Microtus montanus* and *M. pennsylvanicus*, which showed overlap of small numbers into each other's preferred habitats. The former was most abundant in drier areas and the latter in wetter areas. Laboratory studies in this case found that interspecies aggression was greater than intraspecies aggression and the species may well be segregated except for dispersing animals diffusing into less-preferred habitats.

Habitat segregation can be quite sharp giving zonation or continguous allopatry. Miller (1964) reports the example of four species of pocket gophers which form a displacing series in Colorado. *Thomomys talpoides* is the smallest and most tolerant of soil conditions ranging from deep sand to shallow coarse deposits. At the other extreme *Geomys bursarius* is confined to deep fine soils and has obvious fossorial adaptations which reduce its walking ability. These include reduced eyes and ears as well as large claws and splayed forefeet. The other two species are intermediate in size forming a stepped series, but are not mole-like in form. Their distributions may interdigitate where various soil types occur in patches, for example, *Geomys* extends up river margins into the ranges of *T. bottae* and *T. talpoides* and the smaller species is displaced from the deeper soil by the next larger in the series. The smallest species would presumably occupy the entire region in the absence of the others, but the largest species cannot displace the others from the entire range because deep soil is necessary for large burrows and it is too vulnerable when not in a burrow.

These animals are territorial and direct aggression is apparently the cause of displacement of the smaller by the larger. In laboratory studies (Baker, 1974) *T. talpoides* was found to be more aggressive than the next smallest *T. bottae*, in spite of this prediction, and experiments in the field are called for.

A similar case in chipmunks (Sheppard, in Miller, 1967) was investigated by field experiments. *Eutamias amoenus* is consistently larger than *E. minimus* where they occur together, but is less tolerant of open ground between bushes and trees which it uses for escape. *E. amoenus* appears to occupy the more forested areas where their ranges meet and *E. minimus* is then displaced into sagebrush or more alpine habitats. Sheppard removed chipmunks from *E. minimus* and *E. amoenus* habitats and attempted to introduce the opposite species instead. *E. amoenus* did not become established in *E. minimus* habitat, but *E. minimus* survived up to two years in *E. amoenus* habitat. The numbers used were rather low, but do confirm expectation. Direct aggression was investigated in the laboratory and the larger *E. amoenus* won 672 encounters while losing only 23. *E. minimus* may have avoided encounters by remaining in its nestbox for more of the time available.

Two other chipmunk species were investigated by Brown (1971) and show an interestingly different behavioural relationship. *E. dorsalis* is more aggressive and terrestrial than *E. umbrinus*. When trees are sparse *E. dorsalis* can chase *E. umbrinus* away from food and appears to exclude it from Nevada piñon-juniper forest where their ranges meet. At higher altitude the canopy becomes more continuous and *E. umbrinus* escapes through the tree branches along routes which *E. dorsalis* cannot follow.

Trial introductions of species into sites where they do not occur, but which appear suitable, provide another approach to the problem of demonstrating displacement.

Two of the species of planarians worked on by Reynoldson and Bellamy, (1971) are more similar than other possible pairs and clear food differences have not been demonstrated. A 2.5-ha lake in N. Wales apparently suitable for *Polycelis tenuis* contained *P. nigra*, but no *P. tenuis* and to test for exclusion a population of 2500 *P. tenuis* was added experimentally. The introduction failed.

A second experiment was performed in a small artificial pool containing the food animals *Lumbriculus*, *Asellus* and *Potamopyrgus*. The only triclad present was *P. tenuis* and Reynoldson predicted that *Dugesia polychroa* could be introduced because it fed on *Potamopyrgus* while the other flatworm species did not. Introduction of this species was successful. The two *Polycelis* species should not be able to coexist on these foods according to Reynoldson and introduction of *P. nigra* failed.

The causes of these failed introductions are not really known and it seems unlikely that the species are actually identical in exploiting resources. A natural arrival of *P. tenuis* was observed at Llyn Pen-y-parc, a lake used as a reservoir on Anglesey, and this makes an interesting comparison. *P. nigra* was the only triclad species recorded here from 1948–67 in spite of examining 'hundreds of thousands' of specimens and its apparent suitability for all four lake species. *P. tenuis* was found first in 1967 and *P. nigra* declined rapidly in relative abundance by August 1967. *P. tenuis* peaked at 240 per ha in 1968, but then levelled off at 150 per ha replacing *P. nigra* which fell from 160 to 20 per ha and persisted at this level to 1969. This was possibly a stable mixture.

Other factors including selective predation by *Erpobdella* may have had some effect and, as Reynoldson says, this demonstrates the difficulty of using relative distribution and abundance *per se* as evidence of competition. Comparison with other lakes makes these alternatives less likely, but, to test whether the water had changed in some way to cause an independent decline of *P. nigra*, caged populations were suspended in a lake and their survival recorded. No difference in survival or fecundity of the two *Polycelis* species was found.

Cages have been used to test other freshwater animals. Sprules (1972), for example, used them to test the survival of planktonic crustacean species in neighbouring ponds where they had not been recorded in five years of study. Evidence for displacement of *Daphnia rosea* from shallow ponds by *D. pulex* was obtained and similarly for displacement of *Diaptomus coloradensis* by *Di. shoshone*. The actual causes of these displacements were not investigated except to establish that predators ate the larger species in deeper ponds and this was apparently the reason for absence of *D. pulex* and *Di. shoshone* in deep ponds.

5.4   Conclusion

Displacement has been less well documented than coexistence with competition, but this may not reflect the relative frequency of these processes in nature. Coexistence presents a challenge to researchers and the examples where displacement has occurred are no longer available for investigation or are not as obvious. There is ample evidence for the importance of interference in displacement in almost all examples.

References

Baker, A. E. M. (1974). Interspecific aggressive behavior of pocket gophers *Thomomys bottae* and *T. talpoides. Ecology* **55**, 671–3.

Balat, F. (1977). The effect of local suppression of nesting competition of *Passer montanus* on the utilisation of nest boxes by other bird species. *Folia Zool.* **26**, 341–53.

Brian, M. V. (1952). Interaction between ant colonies at an artifical nest site. *Ent. Mon. Mag.* **84**, 8.

Brown, J. H. (1971). Mechanisms of competitive exclusion between two species of chipmunks. *Ecology* **52**, 305–11.

Connell, J. H. (1961). The influence of interspecific competition and other factors on the distribution of the barnacle *Chthalamus stellatus. Ecology* **42**, 710–23.

Davis, J. (1973). Habitat preferences and competition of wintering juncos and golden-crowned sparrows. *Ecology* **54**, 174–80, (The Ecological Society of America).

Elton, C. S. (1958). *The Ecology of Invasions by Animals and Plants.* London: Methuen.

May, R. M. (1973). *Stability and Complexity in Model Ecosystems.* Princeton: Princeton University Press.

Menge, B. A. (1976). Organisation of the New England rocky intertidal community: role of predation, competition, and environmental heterogeneity. *Ecol. Monogr.* **46**, 355–93.

Miller, R. S. (1964). Ecology and distribution of pocket gophers (Geomyidae) in Colorado. *Ecology* **45**, 256–72.

Miller, R. S. (1967). Pattern and process in competition. *Advances Ecol. Res.* **4**, 1–74.

Park, T. (1954). Experimental studies of interspecies competition. II. Temperature, humidity, and competition in two species of *Tribolium. Physiol. Zool.* **27**, 177–238.

Redfield, J. A., Krebs, C. J. and Taitt, M. J. (1977). Competition between *Peromyscus maniculatus* and *Microtus townsendii* in grasslands of coastal British Columbia. *J. Anim. Ecol.* **46**, 607–16.

Reynoldson, T. B. and Bellamy, L. S. (1971). The establishment of interspecific competition in field populations with an example of competition in action between *Polycelis nigra* and *P. tenuis.* In *Proc. Advan. Study Inst., Dynamics of Numbers in Populations*, Oosterbeek, 1970, eds. P. J. den Boer and G. Gradwell. Wageningen: Centre for Agricultural Publishing and Documentation (PUDOC).

Sprules, W. G. (1972). Effects of size selective predation and food competition on high altitude zooplankton communities. *Ecology* **53**, 375–86.

Stoecker, R. E. (1972). Competitive relations between sympatric populations of voles (*Microtus montanus* and *M. pennsylvanicus*). *J. Anim. Ecol.* **41**, 311–29.

6 The consequences of competition in ecological communities

6.1 How many species?

The diversity of life is the main sustaining interest of many biologists and the search for any pattern or general laws to provide an ordered basis for this astounding variety is an inbuilt property of our minds. It should no longer need saying that species are real entities whether taxonomists have described them or not and genetic isolation between species is the main origin and reason for maintenance of such diversity. Exceptional organisms which provide difficulty (*see* Cain, 1954) in species diagnosis do not invalidate these points.

Why, then, are there so many species? This has stimulated much discussion (e.g. Hutchinson, 1959), but one might just as well ask why there are not more species, whether we have some sort of dynamic balance giving a fixed number of species, or if we are observing a process of increasing complexity which will result in far more species given the opportunity?

The reasons for the number of species in any locality are many and various. Not all of them directly involve competition, but act with it under appropriate circumstances so it is worth listing them here:

(1) Severity of physical conditions. Life on earth seems to be adapted to modal conditions and the extreme poles or deserts have few species.

(2) The time available for colonization and speciation. The colonization of Krakatoa (*see* Hesse, Allee and Schmidt, 1951) provides an example of the relative rapidity of the former and the speciation of gammarids in Lake Baikal (Kozhov, 1963) is a convincing example of the importance of time for the evolution of diversity.

(3) Isolation. Geographical isolation of islands such as Easter Island or lakes such as Baikal has presumably contributed to the sparsity of some groups. Simberloff and Wilson (1970) performed a direct test of this by fumigating six islands off the Florida Keys. The island nearest the mainland (2 miles off) reached a much higher equilibrium species number (30–40) than did the furthest island (535 miles and 10–20 spp).

(4) Area. Small islands tend to have lower numbers of species than expected from their position (see MacArthur and Wilson, 1967), perhaps because the extinction rate is higher and a dynamic equilibrium between colonization and extinction determines the actual number present at one time. Any small area

of isolated habitat might be considered an island and Janzen (1968, 1973) discusses the possibility of analysing the fauna of hostplants in terms of island biogeography (*see also* Claridge and Wilson, 1978).

(5) Spatial heterogeneity. Complexity of physical features obviously increases the number of habitats which could be occupied by specialists and further complexity is produced by the organisms themselves. Trees, for example provide opportunity for many different strategies of exploitation from wood-borers to sap-suckers.

(6) Food-chain limits. Predators are usually larger in size and smaller in number than their prey so there must be limits to the number of stages in a food chain (*see* Hutchinson, 1978).

(7) Predator behaviour. If the dominant species, whichever it is, gets eaten more frequently than chance compared with the other possible prey, species richness may be maintained (Chapter 4.13).

(8) Competition. A very large quantity of work has been published which describes the overlap between species in a community and the relevance of this to competition or *vice versa* needs to be discussed.

Communities do appear to be structured. Elton (1927) formalized the concept of a food-web (food cycle) and went on (1946) to consider the role of competition in this structure by investigating the number of species per genus in communities (*see also* Monard, 1920).

One species is unlikely to be the normal prey of another species in the same genus or, in other words, if two species in the same food-web are congeneric they are expected to be in the same trophic level and potentially competing. About 85% of the genera recorded from 55 communities possessed only one species in any one community, but the difficulties of this approach are many. The limits to a community are subjective since they may be subdivided into smaller subcommunities – a wood or one tree or one tree-hole may each be called a community. Generic limits are also subjective, but presumably assessed independently of community ecology. Competition also occurs between distantly related species and the food-web is an oversimplification because species may change status with age and size.

Nevertheless, attention was focused on closely related species although it had been pointed out (e.g. Diver, 1940) that very closely related and ecologically similar species often do occur together. Is there any limit to the similarity of coexisting species?

Stable coexistence when sharing resources depends on there being some difference between the species, but of course differences can be expected between all species. It could be argued (Pontin, 1963) that an infinitesimal difference in the absence of destabilizing interference could result in stable coexistence, especially since disturbing factors would tend to alter the numbers of similar species in a similar manner. This could approach the theoretical limit where two identical species would only be subject to chance displacement of one by the other. May and MacArthur (1972) make a valuable suggestion that in a stochastic system some finite difference is necessary to stabilize coexistence in the face of random processes which might otherwise cause drift to extinction of one of the species. Unfortunately their model is not helpful. Several species are supposed to share a resource continuum such that each uses a normally distributed, overlapping, range (Fig. 6.1). If the distance between the peak frequency of resource used by

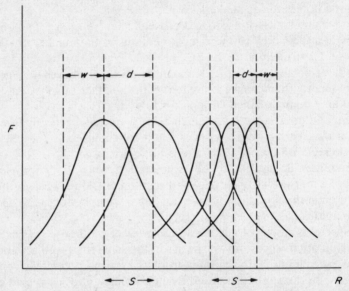

Figure 6.1 The species-packing problem. Suppose the range of resources (R) available could be represented by the abscissa of the graph and they are exploited by two groups of species with the frequencies (F) indicated by the normal curves. Could more species occupy resource space S? The distance d between them may be related to the degree of specialization w, but has w any lower limit? In some of the examples in Chapter 4 the overlap is presumably much greater than that shown here.

neighbouring species pairs in the series is d, and the standard deviation of each species' range of resource is w, then the number of species is predicted to be maximum when d/w approximates to 1. In reality such resource continua have not been discovered and discontinuities or qualitative differences can be expected. If some physical gradient such as temperature is substituted for the x-axis then it will be partitioned by interference and the model does not deal with this vital component of competition. The most serious objection is that the model simply does not do what it is intended to do; there is still no limit set to the number of more specialized (small w) species which can be packed in. d needs to be measured in absolute and not relative terms and this leads us to the most difficult question. Is it realistic to ask a question which needs a reply in terms of general species difference? How does one compare a difference between two tree species with a difference between two mammal species or two ant species? It must be doubted whether we can expect to provide a general predictive theory which specifies measurable minimum difference between coexisting species. But we should not stop trying.

Perhaps the limit has not been reached and we are observing a range of species which is still increasing with more and more complex communities? The countering forces are extinction rate, which may become higher with increasing specialization and smaller numbers of individuals per species, and also interference competition which is even more difficult to measure in comparative terms than exploitation. Dynamic balance may have been reached already, but there is no reason to suppose this is so. Man's interference may well influence the answers to this problem.

6.2 The niche concept

The early usage of the term niche (Grinnell, 1924) meant the characteristic habitat of each species. Elton (1927) clearly included the interrelationships with other species in his concept of a niche which meant the species' rôle in the community. Analogy with a human social community can still be apt. An ecologist saying 'there goes a lion' should be saying something as illuminating as 'there goes the doctor (or teacher)' in a village community, but the function of a doctor can be subdivided so that a larger community might have a pediatrician and a geriatrician as well. A niche, then, in either meaning is a description of the ecology of a species and there is absolutely no justification for supposing that each area has a number of pigeon-holes into which species can be fitted until the community is full.

The most unfortunate result of using the term niche is to predispose the minds of readers into thinking that species occupy exclusive compartments in communities and therefore competition leads to displacement because there is no room for two species in one niche. We have seen already that competition does not lead to displacement in a number of representative examples.

The further idea 'niche overlap' put forward tongue-in-cheek (Pontin, 1963) is of course a self-contradiction which should invalidate the term niche. Clearly, species frequently do have overlapping requirements which may or may not affect their numbers, but if the overlap shows no such effect it is of no importance in this context. All species on earth are sharing its resources so a definition of competition must be restricted to examples where interaction can be demonstrated if this term too is not going to become useless.

Analysis of niches into their component factors, called 'niche dimensions', has become a popular approach to research and we should examine the choice of factors measured to clarify the inadequacy of this method for investigation of competition.

In Fig. 6.2 the species show overlap in both niche dimensions which are represented by the two axes. The dimensions could be a spectrum of food or some habitat dimension such as temperature or height in vegetation (*see* Hutchinson, 1978, for a review). The species overlap less if the dimensions are combined to form areas of exploitation. Further dimensions could be added to give less overlap and the niche then becomes an *n*-dimensional space or hypervolume.

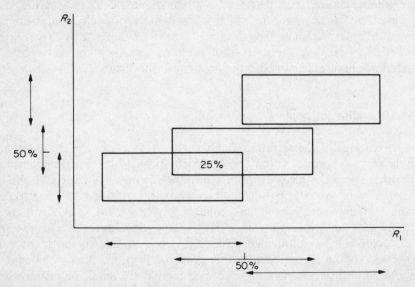

Figure 6.2 The 'niche' overlap problem. Suppose the abscissa represents a range of food resources (R_1) and three species are potentially able to exploit the parts of this range indicated by arrows. Adjacent species overlap by 50%. If the ordinate represents some other environmental factor such as space or time (R_2) which can be divided by the species into zones for exploiting these resources then does species pair 1/2 now overlap by 25% and species pair 2/3 have no overlap? Difficulty arises if the same population of resources extends across the zones of 'niche dimension' two. Of course, if the actual resources taken by the species are measured then other factors have already been allowed for and should not be counted twice by considering them as niche dimensions. The overlap will only determine the severity of competition if the food is scarce enough to increase mortality or reduce fecundity.

How does one choose which factors to include as niche dimensions and, equally important, which ones are to be left out? It is possible to include irrelevant dimensions which give a false impression of no overlap; for example, the height up an oak tree or the time of day for predators feeding on caterpillars. If one species was diurnal and the other noctural, to take an extreme case, they might nevertheless be feeding on the same food population and causing a lower density of it to be available to each other.

It is also possible to omit vital factors and most such studies omit parasites and predators which could render food sharing by their victims unimportant. Interference is also usually omitted and is particularly difficult to measure or incorporate in this sort of treatment, but factors which are unimportant for exploitation may be important in preventing interference.

It would be impracticable to include all possible factors and impossible to know that all had been included.

In theory there is a possible way to select the factors important to competitors. Williamson (1957) argued that competitors have a 'controlling factor' in common and that any species sharing a controlling factor would be competing. A controlling factor is a density-dependent factor which is effective in limiting growth of the population. The food shortage concept of many authors is of course selecting an important factor in precisely this way, but Williamson also included other possible controlling factors such as predation and in this case it may be said that the prey species are competing for the best hiding places. The normal meaning of competition makes this less acceptable to many, but it is ecologically useful to include all cases of reciprocal density reduction under one term.

Unfortunately the term controlling factor has one of the main disadvantages of the term niche. It leads one to divide nature up into non-existent compartments. Suppose that the controlling factor is food shortage and one species takes food items in classes a, b, c and d while the other species takes c, d, e and f it is obvious that all degrees of sharing from the insignificant to complete overlap are possible. Controlling factor overlap would still be preferable to niche overlap, but it is essential to specify what is being shared.

In practice it is difficult to identify controlling factors (*see*, for example, the winter moth study, Varley, Gradwell and Hassell, 1973). It is often easier to show that the species are affecting each others' density by field experiment and it is advisable to do this anyway to be sure that the overlap is ecologically important.

A further problem may arise if and when competition research becomes more sophisticated. Density-independent factors also alter the equilibrium density of populations so it is conceivable that some instances of reciprocal reduction of numbers are the result of shared density-independent factors, but such effects are likely to be of secondary importance.

6.3 Niche dimensions and limits to similarity

In spite of the doubtful relevance of many niche analyses it is possible that some empirical pattern may be shown which is characteristic of coexisting members of a guild. Hutchinson (1959) suggested that the ratio of size of a species to the next largest in the guild as shown by his mammal and bird examples is around 1:1.3. He used linear measurements associated with feeding structures for these and then went on to use the overall length of three European false-waterboatmen (*Corixa*) species for comparison. Horn and May (1977) took the idea further and compared it with Dyar's 'Law' which claims that the successive insect instars also have a ratio of 1:1.3 while Maiorama (1978) attempts to explain the size of this difference by the variation in each class and the need for spacing of the means to avoid overlap.

Can this simple relationship be generally true? Hutchinson's examples range in ratio from 1.09 (*Apodemus sylvaticus/A. flavicollis*) to 1.43 (*Geospiza fortis/G. fuliginosa*) which is rather wide and represents a ratio of bodyweight from 1.3 to nearly 3. The upper limit may well be limited by the examples which are considered (subjectively?) to be in the same guild and the lower limit could be 1.0 from this small number of cases. Abbott, Abbott and Grant (1977) provide much further information about the *Geospiza* species (seed and fruit eating Darwin's finches) and show that the overlap in beak depth of sympatric pairs of species ranges from 1.0 to 2.5. This measurement increases allometrically with body weight and the overlap is related to overlap in diet. Nine out of eighteen pairs of sympatric species had diets overlapping by more than 50%.

The study of waterboatmen (*Buenoa*) in Costa Rica by Gittelman (1975) appears to support the Hutchinson hypothesis since the combinations of the same species found in the same pond suggest a pattern of non-overlap in size. However, the pairs found together had ratios between the length of the most similar species ranging from 1.08 to 1.56. Uetz (1977) caught 520 specimens belonging to 30 species of spiders by pitfall trapping. In this case the species pairs less than 6 mm in length had ratios of length ranging from 1.2 to 1.4 while those more than 6 mm in length were very similar in size with ratios of 1.02 to 1.05. These morphologically similar species showed temporal differences in their activity periods and clearly it is too simple to measure size only. Overlap in the immature stages must also complicate the issue.

The small-mammal fauna of the Sonoran desert was compared with that of the Great Basin desert by Brown (1975) and provides a convincing case for some sort of size structure. The body-weight ratios of pairs of species in a linear sequence of size range from around 1.5 in smaller species to 2.7 in the larger ones (Fig. 6.3) and body weight is well correlated with the seed size taken. Some of the species are bipedal and some are torpid in winter, but there seems no reason to bring other such factors into the discussion.

A very different group of organisms was described using microhabitat width and overlap by Hair and Holmes (1975). The gut helminths of 10 lesser scaup, *Aythya affinis* were distributed along the intestines of their hosts in a series with some overlap between, for example, some adjacent species of the tapeworm genus *Hymenolepis*. The overlap differed in different individual hosts, but was very small in some cases: a very clear illustration of a community structure with all the species sharing the same resource by spatial stratification.

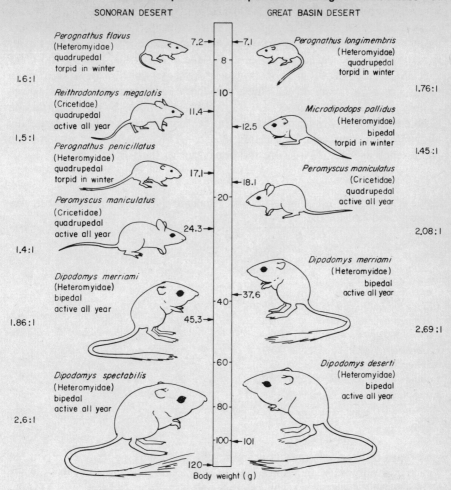

Figure 6.3 Schematic representation of convergence in structure between a six-species community from the Sonoran Desert (Rodeo B, left) and a five-species community from the Great Basin Desert (Dunes 3 and 7, right). Numbers are average weight (g), and size ratios are shown. Note the similarities in body size, form, taxonomic affinity, and other characteristics between species occupying similar positions in each community. Also notice the displacement in body size in *Peromyscus maniculatus* and *D. merriami* (both are larger in the Sonoran Desert) to compensate for the different numbers and sizes of coexisting species. (*From* Brown, 1975.)

Another example of guild analysis is especially interesting because interference is measured. Feinsinger (1976) working on 14 species of coexisting hummingbirds showed that three types of resource variation allowed divergence in exploitation: the species and flower density of the plants, the height of the flowers, and thirdly the time of day related to nectar renewal. The two most abundant species overlapped 17%, no overlap was more than 21% and many of the less-common species showed very little overlap. High flower density tended to cause convergence of the birds and there was no clear specialization of beak length to flower species. All the flowers were of open form. *Amazilia saucerotti* was the most abundant hummingbird species and dominated the rich clumps of flowers. It displaced others by direct aggression (Table 6.1), but it was territorial and far more intraspecific than interspecific encounters were recorded for this species. This should tend to maintain coexistence as well as structure the behaviour of the whole guild.

Table 6.1 Aggressive encounters between hummingbirds when foraging at flowers (From Feinsinger, 1976)

	Total intraspecific encounters	Interspecific encounters with a clear winner	% interspecific wins
Amazillia saucerottei	1361	296	85
Chlorostilbon canivetti	24	84	8.3
Philodice bryantae	96	121	1.7
Colibri thalassinus	21	60	53
Hylocharis eliciae	4	63	39
Eupherusa eximia	25	66	29
Lampornis calolaema	0	29	100
Elvira cupreiceps	14	17	0
Amazillia tzacatl	0	13	69

Perhaps patterns of community structure might emerge more clearly if we considered only those examples with experimental evidence that the species were affecting each other? The examples already discussed (Chapter 4) show either no clear size differences (*Erythroneura*, *Daphnia carinata* agg., *Lasius*) or various weight ratios (starfish 50:1, *Parus* 1.6:1) and no other pattern is yet evident.

There is no reason to deduce from the examples of competition that there are any limits to similarity between coexisting competitors. Limits are imposed by interference where larger size difference increases the likelihood of displacement or the difference is a qualitative one such as toxin production. The structure of guilds is the product of natural selection acting on several generalist species to produce the same number of more-efficient specialists. One can always imagine extra species which could be added to any community with further specialization to follow. Perhaps the very similar competitors discussed will diverge further, are prevented from diverging by their mode of life, or will evolve interference methods to become locally dominant.

References

Abbott, I., Abbott, L. K. and Grant, P. R. (1977). Comparative ecology of Galapagos ground finches *Geospiza*. Evaluation of the importance of floristic diversity and interspecific competition. *Ecol. Monogr.* **47**, 151–84.

Brown, J. H. (1975). In: *Ecology and Evolution of Communities*, pp. 313–41, ed. M. L. Cody and J. M. Diamond. Harvard University: Bellknap Press.

Cain, A. J. (1954). *Animal Species and Their Evolution*. London: Hutchinson.

Claridge, M. F. and Wilson, M. R. (1978). British insects and trees: a study in island biogeography of insect/plant coevolution? *Amer. Naturalist* **112**, 451–6.

Diver, C. (1940). The problem of closely related species living in the same area. In: *The New Systematics*, pp. 303–28, ed. J. S. Huxley. Oxford University Press.

Elton, C. S. (1927). *Animal Ecology*. London: Sidgwick and Jackson.

Elton, C. S. (1946). Competition and the structure of animal communities. *J. Anim. Ecol.* **15**, 54–68.

Feinsinger, P. (1976). Organisation of a tropical guild of nectarivorous birds. *Ecol. Monogr.* **46**, 257–91, (The Ecological Society of America).

Gittelman, S. H. (1975). The ecology of some Costa-Rican backswimmers Hemiptera, Heteroptera: Notonectidae. *Ann. Entomol. Soc. Amer.* **68**, 511–18.

Grinnell, J. (1924). Geography and evolution. *Ecology* **5**, 225–9.

Hair, J. D. and Holmes, J. C. (1975). The usefulness of measures of diversity, niche width and niche overlap in the analysis of helminth communities in waterfowl. *Acta Paras. Polonica* **23**, 253–69.

Hesse, R., Allee, W. C. and Schmidt, K. P. (1951). *Ecological Animal Geography*. New York: Wiley.

Horn, H. S. and May, R. M. (1977). Limits to similarity among coexisting competitors. *Nature* **270**, 660–1.

Hutchinson, G. E. (1959). Homage to Santa Rosalia or why are there so many kinds of animals? *Amer. Naturalist* **93**, 145–59.

Hutchinson, G. E. (1978). *An Introduction to Population Ecology*. New Haven and London: Yale University Press.

Janzen, D. H. (1968). Host plants as islands in evolutionary and contemporary time. *Amer. Naturalist* **102**, 592–5.

Janzen, D. H. (1973). Host plants as islands. II. Competition in evolutionary and contemporary time. *Amer. Naturalist* **107**, 786–90.

Kozhov, M. (1963). Lake Baikal and its life. *Monogr. Biol.* **11**, 1–344.

MacArthur, R. H. and Wilson, E. O. (1967). *The Theory of Island Biogeography*. Princeton University Press.

Maiorama, V. C. (1978). An explanation of ecological and developmental constants. *Nature* **273**, 375–7.

May, R. M. and MacArthur, R. H. (1972). Niche overlap as a function of environmental variability. *Proc. Nat. Acad. Sci. USA* **69**, 1109–13.

Monard, A. (1920). La faune profonde du lac de Neuchâtel. *Bull. Soc. neuchâteloise Sci. Nat.* **44**, 65–236.

Pontin, A. J. (1963). Further considerations of competition and the ecology of the ants *Lasius flavus* (F.) and *L. niger* (L.). *J. Anim. Ecol.* **32**, 565–74.

Simberloff, D. S. and Wilson, E. O. (1970). Experimental zoogeography of islands: a two-year record of colonisation. *Ecology* **51**, 934–7.

Uetz, G. W. (1977). Coexistence in a guild of wandering spiders. *J. Anim. Ecol.* **46,** 531–41.

Varley, G. C., Gradwell, G. R. and Hassell, M. P. (1973). *Insect Population Ecology: an Analytical Approach.* Oxford: Blackwell.

Williamson, M. H. (1957). An elementary theory of interspecific competition. *Nature* **180,** 422–5.

7 Biogeographical consequences of competition

7.1 Competitive release and character divergence

Considered here is an extensive collection of work which compares the ecology of a species when it is within or outside the range of its presumed competitor. Darwin in *The Origin of Species* (1859) pointed out that natural selection leads to divergence of character because 'more living things can be supported in the same area the more they diverge in structure, habits and constitution.' Grant (1972) redefined character displacement as 'the process by which a morphological character state of a species changes under natural selection arising from the presence, in the same environment, of one or more species similar to it ecologically and/or reproductively.' Competitive release is the opposite concept which suggests less specialization, or broadening of the niche, in the absence of a competitor. Most examples are weakened because the absence of a competitor may be the result of an undiscovered factor which also causes the change in the species present. Historical evidence is also lacking and we have the usual problem of investigating evolutionary hypotheses.

Rather than use such examples as evidence of competition it is probably better to regard them as phenomena which may be explained by competition. As well as changes in morphology, changes in genetic constitution, resource utilization, habitat range and density have been described.

Morphological shift

Hutchinson (1959), using skull measurements from Miller (1912), gives a deceptively simple example in European *Mustela*; the stoat and the weasel. The stoat is clearly intermediate in size where it occurs alone in Ireland (Table 7.1) and the inference is that it uses the full range of prey where its competitor is absent so that its size has become adapted to fit. Similarly the weasel in southern Europe where the stoat is absent is larger than it is in Britain where they are sympatric. This highlights the difficulty of interpretation because we are dealing with climatically and faunistically different areas which may well need different adaptation, even if competition were irrelevant and we do not know if it is. The other examples given by Hutchinson and also those given by Brown and Wilson (1956) are heavily critized by Grant (1972),

Table 7.1   Skull size in mustelids when sympatric and allopatric (From Miller, 1912)

		Size	Ratio stoat/weasel	
	Sympatric	Allopatric	Sympatric	Allopatric
♂ Weasel (*Mustela nivalis*)	39.3	42.9 (S. France, Italy)	1.28	1.07
		40.4 (Iberian)		1.14
♂ Stoat (*M. erminea*)	50.4	46.0 (Ireland)		
♀ Weasel	33.6	34.7 (S. France, Italy)	1.34	1.21
		36.0 (Iberian)		1.16
♀ Stoat	45.0	41.9 (Ireland)		
Ratio of ♂/♀ for comparison:		· Weasel	1.16	1.24
				1.12
		Stoat	1.12	1.10

but one more of these is worth quoting here. Wilson (1955) described eight independent characters which could be used to identify the ant *Lasius nearcticus* from the closely related *L. flavus* where they are sympatric in eastern USA. Where *L. flavus* occurs alone its variability is greater and all but two of the diagnostic characters are within its range of variation. It may be doubted whether all the characters are really independent genetically and one wonders just how many species are involved – one, two or three? In this case the shift in morphology is of unknown ecological significance.

A much more elaborate ant example has since been investigated by Davidson (1978). *Veromessor pergandei* is a seed harvesting ant common in the deserts of north and central America. The workers vary in body length from 3.5 to 8.4 mm and each nest normally shows a large range of size in its members. The width of seed brought back to the nests is correlated with size of worker and mandible length was used to compare the size distribution of ants in various localities. Other harvesting ant species occur in the same sites and the variability of the *Veromessor* at bait trays was found to be lower where other species removed bait from the same trays and related to the amount of bait they removed. The smallest size class of *Veromessor* was absent where the smaller *Solenopsis xyloni* was present and with the larger *Pogonomymex* or *Novomyrmex* the *Veromessor* size distribution was skewed toward the smaller end of the range (Fig. 7.1). Presumably there is a selective advantage in producing the most effective worker size for the food available and this availability is altered by other species taking seeds of a size appropriate to them. The potential importance of this example makes it vital to know whether these differences in size range are inherited. They could be produced by predators which select ants of restricted size range and occur most abundantly where there are most suitable ants or perhaps the ants prey

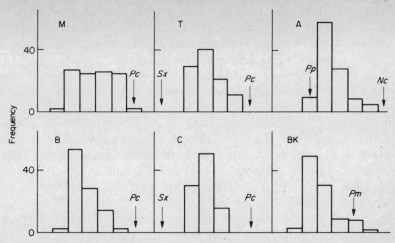

Figure 7.1 Frequency distribution of mandible length of *Veromessor pergandei* from six locations: M, Mojave, California; T, Tacna, Arizona; A, Ajo, Arizona; B, Barstow, California; C, Casa Grande, Arizona; and Bk, Baker, California. The arrows indicate the mandible length of potential competitors at the same locations: *Sx, Solenopsis xyloni*; *Pc, Pogonomyrmex californicus*; *Pp, P. pima*; *Nc, Novomyrmex cockerelli*; and *Pm, P. magnacanthus*. (*From* Davidson, 1978, by permission of the University of Chicago Press.)

on each other. It looks unlikely that local selection by factors other than competition would evolve these differences, but this is precisely what we need to know to establish a real example.

Some genetical evidence for divergence in species of *Acmaea* limpets has been provided by Murphy (1976). Frequencies of allozymes at the leucine aminopeptidase locus were measured for populations of *Acmaea pelta*, *A. digitalis* and *A. scabra* occurring without congeners and also for samples from areas where *A. pelta* coexisted with *A. digitalis* or *A. scabra* coexisted with *A. digitalis*. Frequency differences were found in each case of coexistence compared with that alone, but the effect was asymmetrical: one species affected the other more than the reverse. The cause is not known, but is likely to be direct because *A. pelta* was found to have the same allele frequency for a wide range of habitat and food except where it had other species of acmaeid coexisting.

7.2 Exploitation shift

A number of studies of food range taken by species when sympatric compared with that taken when allopatric have been made.

The two very similar bat species *Myotis auriculus* and *M. evotis* were shown by Black (1974) to prey largely on moths and beetles respectively when sympatric and Husar (1976) investigated their behaviour when allopatric for comparison. They were mist-netted over small areas of water and kept overnight to obtain faecal samples. The proportion of moth scales in the faeces was then used to compare diets. Both species showed less variation in food taken and the main difference in sympatry was interpreted as being 'due to *M. evotis* which becomes a beetle specialist in sympatry.' Whether the difference in foraging behaviour could be caused by direct inteference or whether it is the result of genetic difference is not known. The moth scale content of the faeces is not correlated with relative abundance of moths in insect collections from the same sites so prey selection apparently occurs, but it is difficult to see how dominance of *M. auriculus* over *M. evotis* could produce such differences in selection of prey.

The sunfish worked on by Werner and Hall (1975) were deliberately stocked in ponds so there is presumably no question of natural selection producing the interesting dietary and habitat shifts observed. *Lepomis macrocheirus* feeds in vegetation when alone, but if the water is stocked with *L. cyanellus* it tends to feed more on smaller prey in open water. 44% overlap in diet was estimated in ponds, but if both species were confined to vegetation

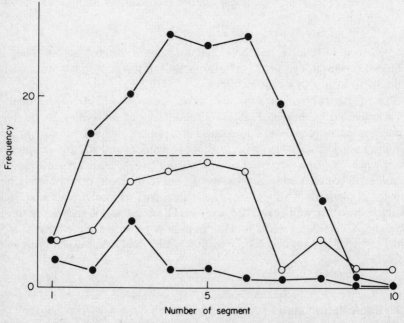

Figure 7.2 Frequency of colonization of *Fucus* segments numbered from the holdfast towards the growing tip (abscissa). Upper curve *Alcyonidium* alone, middle curve *Alcyonidium* with one other bryozoan species, lower curve with any two other species. (*From* O'Connor, Boaden and Seed, 1975, reproduced here by courtesy of *Nature*.)

the overlap was 70%. *L. macrocheirus* showed reduced growth and survival when with *L. cyanellus*, but it was claimed to deal more efficiently with small food so the advantage was not entirely one-sided. Dietary shift is certainly a widespread phenomenon and other examples include the termite species frequency in the diet of *Typhlosaurus* lizards (Huey, Pianka, Egan and Coons, 1974) and the prey size of the sphecid Hymenoptera *Tachysphex terminata* and *T. similis* (Elliott and Kurczewski, 1974).

Habitat shift

A relatively simple example was described by O'Connor, Boaden and Seed (1975) working on a guild of ectoprocts competing for space on fronds of seaweed, *Fucus serratus*, collected in Northern Ireland. The seaweed fronds were divided into zones from holdfast to tip, each zone corresponding to the space between dichotomies in the frond and therefore related to age of segment. Plotting number of segment faces colonized by a species of ectoproct against distance from the holdfast gives a bell-shaped, but non-Gaussian, curve and its width half way up can be used as a measure of habitat width (*see* Fig. 7.2). The commonest species all showed reduced habitat width when others were also present (Table 7.2).

Table 7.2 Effect of number of coexisting species on the habitat breadth of Bryozoa on *Fucus serratus* (From O'Connor, Boaden and Seed, 1975, reproduced here by courtesy of *Nature*.)

No. of other species present	Alcyonidium Breadth	Cases	Electra Breadth	Cases	Flustrellidra Breadth	Cases	Membranipora Breadth	Cases
0	5.9	149	5.1	59	6.8	51	5.0	41
1	4.4	69	2.6	58	6.1	57	1.7	18
2	2.7	18	1.6	20	1.6	20	1.0	2

In contrast, the altitude range of birds in the Andes was interpreted in terms of habitat shift by Terborgh and Weske (1975). Cerros del Sira with 246 species was judged to have 80% of the usual Andean summit zone species missing and 71% of the possible invaders from below extended their range higher than in the 'control' transect in the Cordillera Vilcabamba.

A smaller-scale bird example is shown by the guild of 10 tree-feeding passerines investigated in Sweden by Alerstam, Nilsson and Ulfstrand (1974). In Gotland where crested tit *Parus cristatus*, marsh tit *P. palustris* and willow tit *P. montanus* are absent the coal tit *P. ater* made more intensive use of conifers compared with threecreeper *Certhia familiaris* and goldcrest *Regulus regulus* (Table 7.3) and was apparently also more abundant.

Table 7.3 Feeding stations of passerines in pine trees comparing mainland Sweden where there are extra potential competitors with Gotland (From Alerstam, Nilsson and Ulfstrand, 1974.)

		Frequency (%) on:				
	Trunk		*Branches*		*Twigs and needles*	
	Mainland	*Gotland*	*Mainland*	*Gotland*	*Mainland*	*Gotland*
Coal tit (*Parus ater*)	0	5	15	38	84	68
Goldcrest (*Regulus regulus*)	0	0	8	7	92	93
Treecreeper						
(*Certhia familiaris*)	96	90	4	10	0	0
Crested tit (*P. cristatus*)	12	—	62	—	26	—
Willow tit (*P. montanus*)	15	—	58	—	27	—

7.3 Density shift

Islands have been observed to contain fewer species than the mainland, but a similar or higher total density of individuals. Crowell (1962) found 43% higher density of breeding birds on Bermuda compared with similar scrub habitats in south-eastern USA, but 80% of them were of only three species. Yeaton and Cody (1974) measured the territory sizes of song sparrows *Melospiza melodia morpha* on islands in the Puget Sound district and found a strong positive correlation between territory size and the number of potentially competing species (Fig. 7.3). Overlap in habitat, overlap in vertical stratification and overlap in bill form with feeding behaviour between song sparrows and the other species were estimated. Song sparrows showed an expansion of feeding behaviour into strata where potential competitors were absent and simply replaced them in the community.

Nilsson (1977), working on islands in a Swedish lake, concluded that the combined density of guild members seemed independent of the number of species present in the guild. Turdidae were found to occur at 1.6 pairs per ha however many of the four species were present and similarly *Sylvia* occurred at 1 pair per ha whether there were one, two or three species present. The twig-feeding *Phylloscopus* and *Regulus* were another possible example. The chaffinch (*Fringilla coelebs*) showed a density ten times that of the mainland and this was negatively correlated with the total density of other small birds so it also might provide another example of competitive release. Nilsson points out a number of difficulties in interpreting these island communities. Some birds such as wagtails *Motacilla alba* show high densities on small islands probably because they forage along the shore on emerging insects and the shoreline is relatively longer. Perhaps more serious is the absence of some predators and other sorts of competitor including rodents from islands, but in

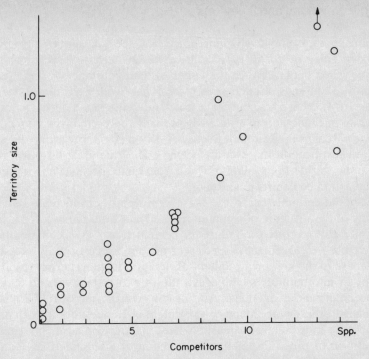

Figure 7.3   The correlation between territory size (acres) (ordinate) of song sparrow and the number of potentially competing species. (*From* Yeaton and Cody, 1974.)

any case it would be valuable to investigate fledgling number to check that the birds were not merely sharing the usual food level out to give smaller rations per pair.

Conclusion

There is a wealth of evidence to show that species change their ecological features where potential competitors are absent. Evidence to show that this is cause and effect is often thin and the ecological consequences to fecundity and density are usually unclear. Also there is usually little discussion of whether the change is the result of natural selection or just adaptable behaviour. Interference is an obvious cause of behavioural change in sympatry and the present topic overlaps with the discussion of displacement in Chapter 5.

Inferences that competition or the lack of it causes competitive release and character divergence cannot be dismissed, indeed it would be surprising if they were wrong, but more experimentation is needed where practicable.

7.4 Introductions and invasions

Invasions of areas previously unoccupied by dispersing species have presumably been going on for as long as life. The Quaternary insect remains, for example, show more constancy of species than constancy of their range as they move thousands of kilometres to keep pace with glaciation changes (*see* Coope, 1978). Man's activities have accelerated the rate of invasion, often by deliberate introduction, to a state where many representatives from different Wallace's Realms now attempt to coexist. Elton (1958) reviewing this subject said with justification that 'we are seeing one of the great historical convulsions in the world's fauna and flora' and went on to speculate about the future and the action ecologists might take.

We should certainly try to use these opportunities to increase our understanding of the structure of communities and many questions could be asked. What are the reasons for failure in introductions? Are communities too rigidly closed to accept new species or are the species simply not suitably preadapted to the new environment? If the introductions become established do they fit into natural assemblages with little disturbance or do they cause displacements and extinctions? Are invaders usually successful because man has changed the environment or the potential competitors in their favour? Perhaps the most relevant question here is: are displacements caused by competition or predation or the introduction of diseases? If competition is the cause of displacement then one can again predict that interference is the key behaviour to look for.

These are even more difficult questions to answer for invaders because of the unpredictability of most arrivals. We rarely know enough about the population dynamics of the species as they existed before invasion and monitoring changes is usually confined to deliberate introductions. Controls are usually not adequate to decide whether an invader or a retreating species is causing the changeover.

The rate of spread of some invaders has been well documented. *Galinsoga ciliata* and *G. parviflora*, cosmopolitan weeds of American origin, showed an exponential increase in distribution records in Britain (*see* Harper, 1977) while the Japanese beetle *Popillia japonica* showed an exponential increase in area colonized in USA with a change to a slower exponential rate after six years. The initial rate was roughly a doubling per year up to 2442 square miles $(6.325 \times 10^3 \, \text{km}^2)$ in 1923. European starlings *Sturnus vulgaris*, which took less than 60 years to spread through the nearctic, and the muskrat *Ondatra zibethica* which spread through the palearctic (augmented by deliberate transfers) were also phenomenally rapid (*see* Elton, 1958).

The communities invaded by these species can hardly have offered much resistance and it is interesting that they have widely differing behaviour and ecological requirements. The weeds are invading disturbed soil by passive transport and the starlings are also invading a man-made habitat, but they displace other hole-nesting birds such as the blue-bird *Sialia sialis* and the flicker *Colaptes auratus* by direct aggression from towns. The Japanese beetle defoliated trees in natural woodland and orchards alike and the muskrat was also apparently easily able to invade natural habitats.

Ants can be among the most spectacular invaders. Many species have become cosmopolitan by transport with foodstuffs on ships. The very versatile *Pheidole megacephala* with a large soldier caste invades houses as well as natural tropical communities and has been particularly dominant on islands. In the Hawaiian group it displaced the previously introduced ant *Solenopsis rufa* and dominated insect life up to 2000 ft (610 m) (Zimmerman, 1948). A later invader, the Argentine ant *Iridomyrmex humilis* removed it completely and also all native ants up to 3000 ft (914 m). The story was similar in Madeira (*see* Haskins, 1945) and the small soft Argentine ant has become the major invading ant in warmer climates. It has two main advantages. It possesses a very effective chemical repellent and it does not inhibit itself by intercolony competition. Colony boundaries seem to be non-existent and interspecific competition is made more severe than intraspecific competition in all conflicts with other ant species. Its rate of spread is not large because the queens are retained by the colonies producing them, but it is continually exploring new ground. Lieberburg, Kranz and Seip (1975) found that *Pheidole* was not completely replaced by *Iridomyrmex* in Bermuda and continues to occur in lowland where it reinvades areas vacated by *Iridomyrmex*.

It is possible to see interference explanations for other interactions including the invasion of oyster beds by the slipper limpet *Crepidula fornicata* (*see* Elton, 1958). The skyscraper method of increasing density and 'overshadowing' of oysters by another filter feeder might be adequate cause of its dominance.

Failure of *Salmo gairdneri* introductions in France where *S. trutta* was also present, but success where it was absent, could also be interference determined (*see* Macan, 1963). *S. trutta* shows highly territorial behaviour in running-water tanks and actively drives off other *Salmo* by direct attack. *S. salar* uses stylized threat postures and grows less well than *S. trutta* when they are together.

Some other freshwater introductions seem to have been successful without observed effects on pre-existing animals. The small operculate snail *Potamopyrgus jenkinsi* (*see* Hunter and Warwick, 1957) is a well known example and another is the flatworm *Dugesia tigrina* which deserves more attention in view of the considerable background of knowledge available on this group (Reynoldson, 1978).

The takeover of California red scale *Aonidiella aurantii* from the yellow scale *A. citrina* in southern California has, on the other hand, been investigated extensively (DeBach, Hendrickson and Rose, 1978). Both have been pests of major importance to citrus growers for a century and are very similar in morphology. They do have ecological differences: both occur on leaves and fruits, but only the red scale feeds on twigs, branches and trunks, the red scale has a much wider range of potential alternative hosts and they are distinguished by the hymenopterous parasites which keep them at a density well below the maximum attainable. The red scale developed cyanide resistance, but this was not the cause of replacement of the yellow scale because the extinctions were more than 15 years after spraying with cyanide had been abandoned. In laboratory tests red scale fecundity is three times that of the yellow and the yellow produces more males and it has a longer generation time. At saturation density on fruits in the laboratory the red scale is clearly and consistently dominant so that yellow scales are eliminated in half a dozen generations compared with 300 generations before field extinctions occurred. Four of the biological control parasites increase the rate of displacement in cultures and only one, *Habrolepis rouxi*, reversed the trend by preferring red scales.

These scale insect species are regarded as ecological 'homologues' by these authors and the displacement is explained by the superior species' ability to exploit all the available resources of the inferior yellow scale. This conclusion is reached because other explanations are discounted. However, we need evidence that the yellow scale would not decline anyway in the absence of the red scale at the same sites. Also an alternative explanation is available. Even if such parasites cannot make specific hosts extinct, the presence of an alternative, but consistently less attractive, host could possibly enable them to extinguish the preferred one.

Two of the parasite species attacking the red scale in California also show interesting geographical displacement (Debach and Sisojevic, 1960). *Aphytis lignanensis* produces at least 10 times as many adults per female parent at 25°C as does *A. chrysomphali* in laboratory trials. It was therefore expected to replace *A. chrysomphali* when introduced as an additional biological control agent. *A. chrysomphali*, however, persisted in the cooler coastal areas and the explanation is apparently that the difference in their reproductive success disappears at 20°C and the shorter generation time of *A. chrysomphali* gives it the advantage.

Schuster and Dean (1976) report another case of an introduced insect parasite of a scale insect (*Antoniae graminis*) reducing the abundance of a competitor. *Anagyrus antoninae* in this case reduced the level of parasitism by *Neodusmetica sangwani* (both encyrtids), but did not displace it. Interference was not detected in laboratory tests of these species and this may well be significant in allowing coexistence.

A further instance from the field of biological control is particularly interesting as a deliberate attempt to displace a pest by a competitor. The principle vector of human filariasis caused by *Wucheria bancrofti* is the mosquito *Aedes polynesensis*. *A. albopictus*, while a known vector of some diseases, is not a vector of filariasis and was introduced (Rosen *et al.*, 1976) on

Taiaro, a polynesian atoll, to see if it would displace *A. polynesensis*. Cultures from three different origins were used and over 1000 females released with males among abundant *A. polynesensis*, but they failed to become established. It is probably expecting too much to get complete displacement, but total failure is disappointing and the reasons for it would be interesting to know. *A. albopictus* males mated with *A. polynesensis* females in the field and this may have contributed to the failure. Further trials under different conditions might be valuable.

Most insect introductions are accidental, but, except for rats, housemice and escapes from captivity, most vertebrate introductions have been deliberate. The spread of several mammal invaders has been associated with reduction in range of potential competitors, but we should not assume cause and effect.

The grey squirrel *Sciurus californensis* in Britain has attracted much attention because of the coincident decline of the red squirrel *S. vulgaris*. The grey spread rapidly from several releases in the late ninteenth century and colonized 10 000 square miles 2.59×10^4 km² by 1930, 20 000 by 1940 and 39 000 by 1960 (*see* Shorten, 1954; Roots, 1976). The red apparently crashed in density before the main spread of the grey and (Middleton, 1930) failed to recover where the grey took over. The range of the red has continued to contract with extension of the grey's range, but it has persisted where the grey has failed to colonize – notably the Isle of Wight. Their diets overlap widely, but the red persists longer in coniferous plantations and the grey is said to be more at home in deciduous woodland which must have been the original habitat of the red in England. Direct combat has been recorded, but some observers deny the importance of this. The grey is about 1.8 times the weight of the red and interference competition must be a likely cause if it can be established that the continued changeover is caused directly by competition.

A parallel problem exists in Europe where the mink *Mustela vison* which escaped from fur farms is taking over the habitat of the otter *Lutra lutra*. The larger otter (2:1 in wt) is declining, but it is recorded to eat mink occasionally. Mink is much more of a Jack-of-all-trades, but their diets converge in winter (Erlinge, 1973) and it may be that the otter is continuing to decline anyway while the mink is merely taking advantage of the situation.

Terman (1975) reviewed the spread of *Sigmodon hispidus* in USA and the coincident declines of density of *Microtus ochrogaster*. *Sigmodon* is clearly dominant in laboratory fights if mature and immatures are not aggressive toward *Microtus*. They have similar diet and may be a simple case of interference displacement.

Man's alteration of the environment is certainly an important factor in providing habitats for invaders and this is perhaps most clearly shown when native species move into new sites provided by our activities. There are many examples, but the spread of the grey kangaroo *Macropus giganteus* into the dryer habitats of the red kangaroo (*M. rufa*) (*see* Richardson, 1975) is an interesting case. Its range has been extended hundreds of miles westward apparently because stock-watering points have enabled it to tolerate previously unsuitable areas. So far the red kangaroo seems to be unaffected.

Distantly related competitors should again not be discounted. Mills and Mark (1977) have investigated the interaction between the takahe *Notornis mantelli* and introduced red deer *Cervus elephas* in New Zealand where they both prefer to eat tussocks of *Chionochloa pallens* and *C. flavescens* rather than other plants. Red deer which eat a wide variety of plants, may be able to destroy the limited preferred food and if the takahe remain less flexible they are in danger of extinction.

New Zealand and other islands with obvious gaps in their fauna have attracted much attention, partly because it has been thought that invasion should be easier if there is room in communities for newcomers. Conservation of their local specialists is also important and introductions have certainly endangered these. Most dangerous invaders have food-chain rather than competitive relationships with the native fauna and flora so there is little evidence that colonization is opposed more by competitors on continents. The failure rate is high in either situation and Roots (1976) records that only 30 of the bird species, out of 150 introduced, became established on the Hawaiian islands compared with 4 out of 30 game birds introduced into North America. Most island invaders have invaded continents as well.

It is possible that displacement is more likely when species without a long history of coadaptation are brought together suddenly, but it is also possible that displacements in other situations have already happened leaving only the stable coexisters for our investigation.

Speculation has been taken much further. Competition has been put forward to explain the supposed superiority of eutherian mammals over marsupials and the latters' smaller range and representation in modern faunas. Other explanations are possible for some modes of life (*see* Cox, 1977), but competition may have played some part. Archer (1976) presents a case for extinction of the Tasmanian wolf (*Thylacinus cynocephalus*) on the Australian mainland by dog. In Tasmania where the dog was not introduced before modern times the thylacine persisted until bounties were given for shooting it in the period 1888 up to a sudden decline in 1909. On mainland Australia dogs were present some 8000 years earlier and thylacines have been absent throughout historic times. As to whether dinosaurs were made extinct by more effective competitors such as mammals – this is going too far into the area of untestable hypothesis.

References

Alerstam, T., Nilsson, S. G. and Ulfstrand, S. (1974). Niche differentiation during winter in woodland birds in southern Sweden and the island of Gotland. *Oikos* **25**, 321–30.

Archer, M. (1974). New information about the quaternary distribution of the thylacine (Marsupialia, Thylacinidae) in Australia. *J. R. Soc. West. Aust.* **57**, 43–50.

Black, H. (1974). A north temperate bat community: structure and prey populations. *J. Mamm.* **55**, 138–57.

Brown, W. L. and Wilson, E. O. (1956). Character displacement. *Syst. Zool.* **5**, 49–64.

Coope, G. R. (1978). Constancy of insect species versus inconstancy of Quaternary environments. *Symp. Roy. Ent. Soc. Lond.* **9**, 176–87. Diversity of Insect Faunas, ed. L. A. Mound and N. Waloff.

Cox, B. (1977). Why marsupials can't win. *Nature* **265**, 14–15.

Crowell, K. L. (1962). Reduced interspecific competition among birds of Bermuda. *Ecology* **43**, 75–88.

Darwin, C. (1859). *The Origin of Species by Means of Natural Selection.* Harvard Facsimile 1st edition, 1964.

Davidson, D. W. (1978). Size variability in the worker caste of a social insect (*Veromessor pergandei* Mayr) as a function of the competitive environment. *Amer. Naturalist* **112**, 523–32.

DeBach, P., Hendrickson, R. M. and Rose, M. (1978). Competitive displacement: Extinction of the yellow scale, *Aonidiella citrina* (Coq.) (Homoptera: Diaspidae), by its ecological homologue, the California red scale, *Aonidiella aurantii* (Mask.) in southern California. *Hilgardia* **46**, 1–35.

DeBach, P. and Sisojevic, P. (1960). Some effects of temperature and competition on the distribution and relative abundance of *Aphytis lingnanensis* and *A. chrysomphali. Ecology* **41**, 153–60.

Elliott, N. B. and Kurczewski, F. E. (1974). Character displacement in *Tachysphex terminatus* and *Tachysphex similis* (Hymenoptera Sphecidae Larrinae). *Ann. Entomol. Soc. Amer.* **67**, 725–7.

Elton, C. S. (1958). *The Ecology of Invasions by Animals and Plants.* London: Methuen.

Erlinge, S. (1972). Interspecific relations between otter *Lutra lutra* and mink *Mustela vision* in Sweden. *Oikos* **23**, 327–35.

Grant, P. R. (1972). Convergent and divergent character displacement. *Biol. J. Linn. Soc.* **4**, 39–68.

Harper, J. L. (1977). *Population Biology of Plants.* London, New York: Academic Press.

Haskins, C. P. (1945). *Of Ants and Men.* London: Allen and Unwin.

Huey, R. B., Pianka, E. R., Egan, M. E. and Coons, L. W. (1974). Ecological shifts in sympatry: Kalahari fossorial lizards (*Typhlosaurus*). *Ecology* **55**, 304–16.

Hunter, W. R. and Warwick, T. (1957). Records of *Potamopyrgus jenkinsi* (Smith) in Scottish fresh waters over fifty years (1906–56). *Proc. Roy. Soc. Edinb. B* **66**, 360–73.

Husar, S. L. (1976). Behavioral character displacement evidence of food partitioning in insectivorous bats. *J. Mamm.* **57**, 331–8.

Hutchinson, G. E. (1959). Homage to Santa Rosalia or why are there so many kinds of animals? *Amer. Naturalist* **93**, 145–59.

Lieberburg, I., Kranz, P. M. and Seip, A. (1975). Bermudan ants revisited: the status and interaction of *Pheidole megacephala* and *Iridomyrmex humilis. Ecology* **56**, 473–8.

Macan, T. T. (1963). *Freshwater Ecology*. London: Longmans.

Middleton, A. D. (1930). The ecology of the American grey squirrel (*Sciurus carolinensis* Gmelin) in the British Isles. *Proc. Zool. Soc. Lond.*, 809–43.

Miller, G. S. (1912). *Catalogue of the Mammals of Western Europe. XV.* London: British Museum.

Mills, J. A. and Mark, A. F. (1977). Food preferences of takahe in Fiordland National Park, New Zealand, and the effect of competition from introduced red deer. *J. Anim. Ecol.* **46**, 939–58.

Murphy, P. G. (1976). Electrophoretic evidence that selection reduces ecological overlap in marine limpets. *Nature* **261**, 228–30.

Nilsson, S. G. (1977). Density compensation among birds breeding on small islands in a south Swedish lake. *Oikos* **28**, 170–6.

O'Connor, R. J., Boaden, P. J. S. and Seed, R. (1975). Niche breadth in Bryozoa as a test of competition theory. *Nature* **256**, 307–9.

Richardson, B. J. (1975). r and K selection in kangaroos. *Nature* **255**, 323–4.

Reynoldson, T. B. (1978). *A Key to British Species of Freshwater Tricalds*. Freshwater Biol. Assoc. Sci. Pub. 23.

Roots, C. (1976). *Animal Invaders*. London, Vancouver: David and Charles.

Rosen, L., Rozeboom, L. E., Reeves, W. C., Saugrain, J. and Gubler, D. J. (1976). A field trial of competitive displacement of *Aedes polynesiensis* by *Aedes albopictus* on a pacific atoll. *Am. J. Trop. Med. Hyg.* **25**, 906–13.

Schuster, M. F. and Dean, H. A. (1976). Competitive displacement of *Anagyrus antoninae* (Hymenoptera: Encyrtidae) by its ecological homologue *Neodusmetia sangwani* (Hymenoptera: Encyrtidae). *Entomophaga* **21**, 127–30.

Shorten, M. (1954). *Squirrels*. London: Collins.

Terborgh, J. and Weske, J. S. (1975). The role of competition in the distribution of Andean birds. *Ecology* **56**, 562–76.

Terman, M. R. (1974). Behavioral interactions between *Microtus* and *Sigmodon*: a model for competitive exclusion. *J. Mamm.* **55**, 705–19.

Werner, E. E. and Hall, D. J. (1975). Niche shifts in sunfishes: experimental evidence and significance. *Science* **191**, 404–6.

Wilson, E. O. (1955). *A Monographic Revision of Ant Genus Lasius*. Bull. Mus. Comp. Zool.

Yeaton, R. I. and Cody, M. L. (1974). Competitive release in island song sparrow populations. *Theor. Pop. Biol.* **5**, 42–58.

Zimmerman, E. C. (1948). *Insects of Hawaii*. Honolulu: University of Hawaii Press.

Index